Sustainable Practices

Climate change is widely agreed to be one the greatest challenges facing society today. Mitigating and adapting to it is certain to require new ways of living. Thus far efforts to promote less resource-intensive habits and routines have centred on typically limited understandings of individual agency, choice and change. This book shows how much more the social sciences have to offer.

The contributors to *Sustainable Practices: Social Theory and Climate Change* come from different disciplines – sociology, geography, economics and philosophy – but are alike in taking social theories of practice as a common point of reference. This volume explores questions which arise from this distinctive and fresh approach:

- how do practices and material elements circulate and intersect?
- how do complex infrastructures and systems form and break apart?
- how does the reproduction of social practice sustain related patterns of inequality and injustice?

This collection shows how social theories of practice can help us understand what societal transitions towards sustainability might involve, and how they might be achieved. It will be of interest to students and researchers in sociology, environmental studies, geography, philosophy and economics, and to policy makers and advisors working in this field.

Elizabeth Shove is Professor of Sociology at Lancaster University and held an ESRC climate change leadership fellowship on Transitions in Practice. Recent publications include *The Dynamics of Social Practice: everyday life and how it changes*, with Mika Pantzar and Matt Watson (Sage, 2012).

Nicola Spurling is Research Associate in the Sustainable Practices Research Group at Manchester University. Her research explores how social practices change, focusing on intersections of policy, institutions and individual biographies.

Routledge advances in sociology

1. **Virtual Globalization**
   Virtual spaces/tourist spaces
   Edited by David Holmes

2. **The Criminal Spectre in Law, Literature and Aesthetics**
   Peter Hutchings

3. **Immigrants and National Identity in Europe**
   Anna Triandafyllidou

4. **Constructing Risk and Safety in Technological Practice**
   Edited by Jane Summerton and Boel Berner

5. **Europeanisation, National Identities and Migration**
   Changes in boundary constructions between Western and Eastern Europe
   Willfried Spohn and Anna Triandafyllidou

6. **Language, Identity and Conflict**
   A comparative study of language in ethnic conflict in Europe and Eurasia
   Diarmait Mac Giolla Chríost

7. **Immigrant Life in the US**
   Multi-disciplinary perspectives
   Edited by Donna R. Gabaccia and Colin Wayne Leach

8. **Rave Culture and Religion**
   Edited by Graham St. John

9. **Creation and Returns of Social Capital**
   A new research program
   Edited by Henk Flap and Beate Völker

10. **Self-Care**
    Embodiment, personal autonomy and the shaping of health consciousness
    Christopher Ziguras

11. **Mechanisms of Cooperation**
    Werner Raub and Jeroen Weesie

12. **After the Bell**
    Educational success, public policy and family background
    Edited by Dalton Conley and Karen Albright

13. **Youth Crime and Youth Culture in the Inner City**
    Bill Sanders

14. **Emotions and Social Movements**
    Edited by Helena Flam and Debra King

15. **Globalization, Uncertainty and Youth in Society**
    Edited by Hans-Peter Blossfeld, Erik Klijzing, Melinda Mills and Karin Kurz

16 **Love, Heterosexuality and Society**
Paul Johnson

17 **Agricultural Governance**
Globalization and the new politics of regulation
Edited by Vaughan Higgins and Geoffrey Lawrence

18 **Challenging Hegemonic Masculinity**
Richard Howson

19 **Social Isolation in Modern Society**
Roelof Hortulanus, Anja Machielse and Ludwien Meeuwesen

20 **Weber and the Persistence of Religion**
Social theory, capitalism and the sublime
Joseph W. H. Lough

21 **Globalization, Uncertainty and Late Careers in Society**
Edited by Hans-Peter Blossfeld, Sandra Buchholz and Dirk Hofäcker

22 **Bourdieu's Politics**
Problems and possibilities
Jeremy F. Lane

23 **Media Bias in Reporting Social Research?**
The case of reviewing ethnic inequalities in education
Martyn Hammersley

24 **A General Theory of Emotions and Social Life**
Warren D. TenHouten

25 **Sociology, Religion and Grace**
Arpad Szakolczai

26 **Youth Cultures**
Scenes, subcultures and tribes
Edited by Paul Hodkinson and Wolfgang Deicke

27 **The Obituary as Collective Memory**
Bridget Fowler

28 **Tocqueville's Virus**
Utopia and dystopia in Western social and political thought
Mark Featherstone

29 **Jewish Eating and Identity Through the Ages**
David Kraemer

30 **The Institutionalization of Social Welfare**
A study of medicalizing management
Mikael Holmqvist

31 **The Role of Religion in Modern Societies**
Edited by Detlef Pollack and Daniel V. A. Olson

32 **Sex Research and Sex Therapy**
A sociological analysis of Masters and Johnson
Ross Morrow

33 **A Crisis of Waste?**
Understanding the rubbish society
Martin O'Brien

34 **Globalization and Transformations of Local Socioeconomic Practices**
Edited by Ulrike Schuerkens

35 **The Culture of Welfare Markets**
The international recasting of pension and care systems
Ingo Bode

36 **Cohabitation, Family and Society**
Tiziana Nazio

37 **Latin America and Contemporary Modernity**
A sociological interpretation
José Maurízio Domingues

38 **Exploring the Networked Worlds of Popular Music**
Milieu cultures
Peter Webb

39 **The Cultural Significance of the Child Star**
Jane O'Connor

40 **European Integration as an Elite Process**
The failure of a dream?
Max Haller

41 **Queer Political Performance and Protest**
Benjamin Shepard

42 **Cosmopolitan Spaces**
Europe, globalization, theory
Chris Rumford

43 **Contexts of Social Capital**
Social networks in communities, markets and organizations
Edited by Ray-May Hsung, Nan Lin, and Ronald Breiger

44 **Feminism, Domesticity and Popular Culture**
Edited by Stacy Gillis and Joanne Hollows

45 **Changing Relationships**
Edited by Malcolm Brynin and John Ermisch

46 **Formal and Informal Work**
The hidden work regime in Europe
Edited by Birgit Pfau-Effinger, Lluis Flaquer, and Per H. Jensen

47 **Interpreting Human Rights**
Social science perspectives
Edited by Rhiannon Morgan and Bryan S. Turner

48 **Club Cultures**
Boundaries, identities and otherness
Silvia Rief

49 **Eastern European Immigrant Families**
Mihaela Robila

50 **People and Societies**
Rom Harré and designing the social sciences
Luk van Langenhove

51 **Legislating Creativity**
The intersections of art and politics
Dustin Kidd

52 **Youth in Contemporary Europe**
Edited by Jeremy Leaman and Martha Wörsching

53 **Globalization and Transformations of Social Inequality**
Edited by Ulrike Schuerkens

54 **Twentieth Century Music and the Question of Modernity**
Eduardo De La Fuente

55 **The American Surfer**
Radical culture and capitalism
Kristin Lawler

56 **Religion and Social Problems**
Edited by Titus Hjelm

57 **Play, Creativity, and Social Movements**
If I can't dance, it's not my revolution
Benjamin Shepard

58 **Undocumented Workers' Transitions**
Legal status, migration, and work in Europe
Sonia McKay, Eugenia Markova and Anna Paraskevopoulou

59 **The Marketing of War in the Age of Neo-Militarism**
Edited by Kostas Gouliamos and Christos Kassimeris

60 **Neoliberalism and the Global Restructuring of Knowledge and Education**
Steven C. Ward

61 **Social Theory in Contemporary Asia**
Ann Brooks

62 **Foundations of Critical Media and Information Studies**
Christian Fuchs

63 **A Companion to Life Course Studies**
The social and historical context of the British birth cohort studies
Michael Wadsworth and John Bynner

64 **Understanding Russianness**
Risto Alapuro, Arto Mustajoki and Pekka Pesonen

65 **Understanding Religious Ritual**
Theoretical approaches and innovations
John Hoffmann

66 **Online Gaming in Context**
The social and cultural significance of online games
Garry Crawford, Victoria K. Gosling and Ben Light

67 **Contested Citizenship in East Asia**
Developmental politics, national unity, and globalization
Kyung-Sup Chang and Bryan S. Turner

68 **Agency without Actors?**
New approaches to collective action
Edited by Jan-Hendrik Passoth, Birgit Peuker and Michael Schillmeier

69 **The Neighborhood in the Internet**
Design research projects in community informatics
John M. Carroll

70 **Managing Overflow in Affluent Societies**
Edited by Barbara Czarniawska and Orvar Löfgren

71 **Refugee Women**
Beyond gender versus culture
Leah Bassel

72 **Socioeconomic Outcomes of the Global Financial Crisis**
Theoretical discussion and empirical case studies
Edited by Ulrike Schuerkens

73 **Migration in the 21st Century**
Political economy and ethnography
Edited by Pauline Gardiner Barber and Winnie Lem

74 **Ulrich Beck**
An introduction to the theory of second modernity and the risk society
Mads P. Sørensen and Allan Christiansen

75 **The International Recording Industries**
Edited by Lee Marshall

76 **Ethnographic Research in the Construction Industry**
Edited by Sarah Pink, Dylan Tutt and Andrew Dainty

77 **Routledge Companion to Contemporary Japanese Social Theory**
From individualization to globalization in Japan today
Edited by Anthony Elliott, Masataka Katagiri and Atsushi Sawai

78 **Immigrant Adaptation in Multi-Ethnic Societies**
Canada, Taiwan, and the United States
Edited by Eric Fong, Lan-Hung Nora Chiang and Nancy Denton

79 **Cultural Capital, Identity, and Social Mobility**
The life course of working-class university graduates
Mick Matthys

80 **Speaking for Animals**
Animal autobiographical writing
Edited by Margo DeMello

81 **Healthy Aging in Sociocultural Context**
Edited by Andrew E. Scharlach and Kazumi Hoshino

82 **Touring Poverty**
Bianca Freire-Medeiros

83 **Life Course Perspectives on Military Service**
Edited by Janet M. Wilmoth and Andrew S. London

84 **Innovation in Socio-Cultural Context**
Edited by Frane Adam and Hans Westlund

85 **Youth, Arts and Education**
Reassembling subjectivity through affect
Anna Hickey-Moody

86 **The Capitalist Personality**
Face-to-face sociality and economic change in the post-Communist world
Christopher S. Swader

87 **The Culture of Enterprise in Neoliberalism**
Specters of entrepreneurship
Tomas Marttila

88 **Islamophobia in the West**
Measuring and explaining
individual attitudes
Marc Helbling

89 **The Challenges of Being a Rural Gay Man**
Coping with stigma
Deborah Bray Preston and Anthony R. D'Augelli

90 **Global Justice Activism and Policy Reform in Europe**
Understanding when change happens
Edited by Peter Utting, Mario Pianta and Anne Ellersiek

91 **Sociology of the Visual Sphere**
Edited by Regev Nathansohn and Dennis Zuev

92 **Solidarity in Individualized Societies**
Recognition, justice and good judgement
Søren Juul

93 **Heritage in the Digital Era**
Cinematic tourism and the activist cause
Rodanthi Tzanelli

94 **Generation, Discourse, and Social Change**
Karen Foster

95 **Sustainable Practices**
Social theory and climate change
Edited by Elizabeth Shove and Nicola Spurling

Sustainable Practices
Social theory and climate change

Edited by Elizabeth Shove and Nicola Spurling

Routledge
Taylor & Francis Group

LONDON AND NEW YORK

First published 2013
by Routledge
2 Park Square, Milton Park, Abingdon, Oxfordshire OX14 4RN

Simultaneously published in the USA and Canada
by Routledge
711 Third Avenue, New York, NY 10017

First issued in paperback 2014

Routledge is an imprint of the Taylor and Francis Group, an informa business

© 2013 Elizabeth Shove and Nicola Spurling

The right of the editors to be identified as the authors of the editorial material, and of the authors for their individual chapters, has been asserted in accordance with sections 77 and 78 of the Copyright, Designs and Patents Act 1988.

All rights reserved. No part of this book may be reprinted or reproduced or utilised in any form or by any electronic, mechanical, or other means, now known or hereafter invented, including photocopying and recording, or in any information storage or retrieval system, without permission in writing from the publishers.

Trademark notice: Product or corporate names may be trademarks or registered trademarks, and are used only for identification and explanation without intent to infringe.

British Library Cataloguing in Publication Data
A catalogue record for this book is available from the British Library

Library of Congress Cataloging in Publication Data
Sustainable practices : social theory and climate change / edited by Elizabeth Shove and Nicola Spurling.
   p. cm. – (Routledge advances in sociology ; 95)
   Includes bibliographical references and index.
   1. Environmentalism–Social aspects. 2. Climatic changes–Social aspects. 3. Sustainable development–Social aspects. 4. Environmental policy–Social aspects. I. Shove, Elizabeth, 1959– II. Spurling, Nicola.
   GE195.S874 2012
   304.2'5–dc23   2012036628

ISBN 978-0-415-54065-0 (hbk)
ISBN 978-1-138-84715-6 (pbk)
ISBN 978-0-203-07105-2 (ebk)

Typeset in Times
by Wearset Ltd, Boldon, Tyne and Wear

Contents

List of contributors		xiii
Acknowledgements		xv

1 **Sustainable practices: social theory and climate change** 1
ELIZABETH SHOVE AND NICOLA SPURLING

PART I
How are practices defined and how do they change? 15

2 **What sort of a practice is eating?** 17
ALAN WARDE

3 **The edge of change: on the emergence, persistence and dissolution of practices** 31
THEODORE SCHATZKI

PART II
The materials of practice 47

4 **Transitions in the wrong direction? Digital technologies and daily life** 49
INGE ROPKE AND TOKE H. CHRISTENSEN

5 **Mundane materials at work: paper in practice** 69
SARI YLI-KAUHALUOMA, MIKA PANTZAR AND SAMMY TOYOKI

PART III
Sharing and circulation 87

6 Practices, movement and circulation: implications for sustainability 89
ALLISON HUI

7 Sharing conventions: communities of practice and thermal comfort 103
RUSSELL HITCHINGS

PART IV
Relations between practices 115

8 Building future systems of velomobility 117
MATT WATSON

9 The making of electric cycling 132
JULIEN McHARDY

10 Extended bodies and the geometry of practices 146
GRÉGOIRE WALLENBORN

PART V
Sustainability, inequality and power 165

11 Power, sustainability and well being: an outsider's view 167
ANDREW SAYER

12 Inequality, sustainability and capability: locating justice in social practice 181
GORDON WALKER

Index 197

Contributors

Toke H. Christensen is Researcher at the Danish Building Research Institute, Aalborg University. He studies the relationship between everyday life, energy use and sustainability with a particular focus on households' use of technologies (including ICT and consumer electronics).

Russell Hitchings lectures in Human Geography at University College London. His research largely considers how different social groups relate to elements of the natural world, with recent projects focusing on seasonality and the experience of different indoor and outdoor environments.

Allison Hui is a Postdoctoral Research Fellow at the David C. Lam Institute for East-West Studies, Hong Kong Baptist University. Her research on everyday practices and travel has been published in Tourist Studies and the Journal of Consumer Culture.

Julien McHardy trained as a designer and is now a PhD student at Lancaster University's Sociology Department. His research on the 'Multiple making of electric bikes', examines different sites in which the discourses, policies, components and practitioners/users of electric cycling are produced.

Mika Pantzar is a Research Professor at the National Consumer Research Centre in Helsinki. He has published articles and books on consumer research, design and technology studies, the rhetoric of economic policy, future studies and systems research.

Inge Røpke is Professor in Ecological Economics at Aalborg University's campus in Copenhagen. She has written about the development of modern ecological economics and has specialised in consumption and environment, technology and everyday life with a particular focus on ICT.

Andrew Sayer is Professor of Social Theory and Political Economy at Lancaster University. His main interest is in inequality and his most recent book is *Why Things Matter to People: Social Science, Values and Ethical Life* (Cambridge University Press, 2011).

Theodore Schatzki is Professor of Philosophy and Geography and Senior Associate Dean of Arts and Sciences at the University of Kentucky. Among his books are *Social Practices* (1996), *The Site of the Social* (2002), and *The Timespace of Human Activity* (2010).

Elizabeth Shove is Professor of Sociology at Lancaster University and held an ESRC climate change leadership fellowship on Transitions in Practice. Recent publications include *The Dynamics of Social Practice: everyday life and how it changes*, with Mika Pantzar and Matt Watson (Sage, 2012).

Nicola Spurling is Research Associate in the Sustainable Practices Research Group at Manchester University. Her research explores how social practices change, focusing on intersections of policy, institutions and individual biographies.

Sammy Toyoki is Assistant Professor in Marketing at the Aalto University, School of Economics. He has a PhD from Warwick University, and is interested in consumption, qualitative methods, narrative, identity and spatiality in consumer culture and social theory.

Gordon Walker is Professor at Lancaster Environment Centre, Lancaster University. His research has focused on environmental justice; sociotechnical change and transitions; and the social dimensions of sustainable energy technologies. He has recently published *Environmental Justice: Concepts, Evidence and Politics* (Routledge, 2012).

Grégoire Wallenborn, physicist and philosopher, works at the Centre for Studies on Sustainable Development (IGEAT-Free University of Brussels). He coordinates different projects on household energy consumption including scenarios of transformation of consumption practices and aspects of social inequalities.

Alan Warde is Professor of Sociology at the University of Manchester and Visiting Professor at the University of Helsinki Collegium for Advanced Studies (2010–2012). He has written widely on the sociology of food, consumption and social theories of practice.

Matt Watson is a Lecturer in Human Geography at the University of Sheffield. Current research projects deal with issues of food and energy, framed and developed with reference to theories of practice, science and technology studies, and concepts of governance.

Sari Yli-Kauhaluoma is a research fellow in Organization and Management at the Aalto University, School of Economics. She has written about temporal aspects of innovation processes and has published in *Time and Society*.

Acknowledgements

The symposium which led to this book was funded by Elizabeth Shove's ESRC Climate Change Leadership Fellowship award no. RES-066–27–0015. Thanks to all the presenters and discussants for their valuable contributions to the Symposium: Climate Change and Transitions in Practice, Lancaster University, July 2010.

We are grateful to Conde Nast for permission to use the cartoon in Chapter 11.

1 Sustainable practices
Social theory and climate change

Elizabeth Shove and Nicola Spurling

Introduction

This book is informed by three propositions: one is that consumption is usefully understood as an outcome of practice: people consume objects, resources and services not for their own sake but in the course of accomplishing social practices (Warde 2005). The second is that mitigating and adapting to climate change is sure to require different patterns both of consumption and daily life. In short, the challenge is one of imagining and realising versions of normal life that fit within the envelope of sustainability and that are resilient, adaptable and fair. Since this implies a substantial, systemic transition in what people *do* – in how they move around, what they eat, and how they spend their time – the third proposition is that social theories of practice provide an important intellectual resource for understanding and perhaps establishing social, institutional and infrastructural conditions in which much less resource intensive ways of life might take hold.

According to the World Wildlife Fund, the resources needed to maintain Western European habits currently exceed the earth's capacity by a factor of three. This has not always been so. In the 1970s, on average, European levels of consumption remained within limits which our planet could sustain. How ordinary ways of life have become so resource intensive in such a short space of time is one of the puzzles that runs through this book. There are different ways of approaching this topic. In popular and policy discourse it is usual to explain such changes as outcomes of individual choice. From this perspective, moving towards a more sustainable society depends on helping people to make better choices, for example by understanding the environmental impact of what they do. Ecological and carbon footprint calculators[1] are designed with this in mind. These calculators sum up the consequences of an individual's diet, their mobility and the way they heat and light their home, presenting the results in simple graphs and figures. By adjusting responses to the calculator's questions, people can quickly see how much difference they could make to the size of their ecological footprint by using the car less, being vegetarian, increasing the number of people who share the same home or taking holidays close to home.

Tools like this draw attention to contemporary expectations of 'normal life' (e.g. car ownership, number of people per household, number of holidays per year, etc.), demonstrating just how embedded and ordinary resource intensive lifestyles have become. Can individuals simply choose to reconfigure their lives in ways that meet the calculator's demands? What other aspects of daily life would have to change? What would it mean to abandon the car, cut domestic energy consumption by a half or a third, move into a smaller home and share that space with others? In what sense are substantially more sustainable ways of living plausible given present transport systems, buildings and social conventions? Responses are sure to vary from case to case but the general pattern is clear. Bringing individual carbon or ecological footprints back into the range of 'one planet' is likely to require a radical redefinition of what counts as normal social practice, and of the institutions and infrastructures on which these arrangements depend.

This might seem like a massive, perhaps insurmountable, challenge but if we look to recent history or take note of contemporary variation, it is obvious that there is no single template to which daily life conforms. The range of social practices that constitute seemingly essential aspects of contemporary living are contingent and constantly shifting. For example, diets and habits of personal hygiene (showering, bathing, etc.) are on average quite unlike those that pertained fifty years ago, and there have been correspondingly significant developments in the technologies, meanings and competences involved. Taking a slightly broader view the total range of practices that constitute social life has also changed: what is normal today has not always been so and, as such, there is no reason to suppose that currently familiar arrangements will stay the same for very long. It is reasonable to expect transitions in the array of practices that constitute social life and the resources they require. From an environmental point of view, the question is whether social practices might develop in directions that lead, en masse, to a spectacular reduction in collective ecological and carbon footprints. Can we imagine how such change might come about, and if so, what further costs and consequences might follow?

In designing a book that addresses these questions, we focus on understanding how contemporary patterns of consumption come to be as they are, and how they might change – a project to which the social sciences have much to contribute. In embarking on this task, we make a number of broad assumptions about what sustainability means and how sustainable practice might be defined. Reducing resource use to a level that can be maintained by future generations seems like a reasonable goal, and is one that is addressed in several chapters. However, contributors do not imagine the existence of a single universally agreed template of what constitutes a lower carbon or more sustainable society. Other possible ambitions include those of scaling back carbon and other greenhouse gas emissions (with a focus on mitigating climate change), minimising the rate at which finite fossil fuels are depleted, preserving biodiversity, and addressing global inequalities in consumption and well being.

Voß and colleagues argue for an approach that 'abandons the assumption of "one" adequate problem framing, "one" true prognosis of consequences, and "one" best way to go' ... as if this ... 'could be identified in an objective manner from a neutral, supervisory outlook on the (social–ecological) system as a whole' (Voß and Bornemann 2011). Most, but perhaps not all, contributors to this book share something of the same reflexive stance. In so far as interpretations and definitions of sustainable goals come into view they are, for the most part, taken to be provisional and historically specific.

Theories of consumption have shifted over the past decade as attention has turned from individual consumers to the cultural, material and economic structuring of consumption (Cohen and Murphy 2001; Gronow and Warde 2001; Shove 2003; Southerton *et al.* 2004; van Vliet *et al.* 2005; Spaargaren 2011). Along the way, this literature has questioned behavioural representations of individual choice, underlined the environmental implications of ordinary consumption and argued for a focus on the services (e.g. lighting, heating, laundering, etc.) that resources make possible. The related conclusion that services are implicated in the reproduction of everyday life represents an important step, but more is required to explain how they evolve. In concentrating on these questions – how do more and less sustainable practices become established, and how do they diffuse? – this book moves the agenda on, and moves it in a direction that exploits recent developments in social theory. It does so by placing 'practices' centre-stage.

The Practice Turn in Contemporary Theory (Schatzki *et al.* 2001) signalled renewed interest in theories of practice. The strategy of taking social practices, ordered across space and time, as the focus of enquiry sets such approaches apart from individualist and structuralist modes of thinking (Giddens 1984) and has further consequences for how processes of social reproduction and transformation are conceptualised. For example, rather than seeing change in the resource intensity of daily life as an outcome of individual choice, or of seemingly external social and economic forces, it makes sense to ask about how social practices evolve, and what this means for the use of energy, water and other natural resources. There is no one theory of practice and no such thing as 'a' practice approach, but in developing different aspects of the 'practice turn', and exploring its implications for sustainability, contributors to this collection address a series of related questions: how are practices defined and how do they change? What are the material elements of which practices are composed? How do practices circulate and travel, and how do they relate to each other? How can we understand power within systems of practice, and how and by whom are matters of value, including interpretations of well being and sustainability, reproduced and contested? In responding to these questions, authors make use of different disciplinary traditions, drawing on sociology, geography, science studies, economics and philosophy, and introduce empirical research on cycling, heating, cooling and eating. In combination, the result is an exploratory venture in which new lines of enquiry are opened up, and different threads developed: it is around these that the five parts of the book are organised.

How are practices defined and how do they change?

The first two chapters deal with fundamental problems of definition and change. If practices are to figure as the basic unit of social enquiry, how are they to be recognised and known? Second, if social practices underpin consumption, how is it that new forms come into being and others disappear? In responding to the first question, Alan Warde's discussion 'What sort of a practice is eating?', draws attention to the importance of formalisation and coordination. Some practices are relatively simple to spot in that they are bound and delimited by shared, somewhat formalised, descriptions, prescriptions and definitions of proper performance. These are the easy cases in that there is likely to be a readily observable trail of relatively unambiguous indicators – documents, rules and guides – demonstrating that a practice is 'out there', existing across space and time, and figuring as a recognisable entity that people can join, defect from or resist.

In trying to make sense of other more troublesome examples, like eating, Warde turns from questions of bounding (where does one practice end and another begin?) to matters of linkage and intersection (how are practices coordinated?). In moving into this territory, and in writing about compound practices, he brings new topics to the fore: in particular, what are the threads of interdependence that hold constellations and complexes of practice provisionally in place and how and why do these threads tighten and slacken? Rather than fretting about whether eating is, or is not, a practice in some absolute sense, the message and the practical methodological advice is to identify and compare stronger and weaker forms of coordination between components like those of shopping, preparing and consuming. This is useful guidance for those interested in analysing the trajectories of compound practices over time. It is also useful in thinking about how practices emerge, persist and disappear, this being the focus of Theodore Schatzki's contribution entitled 'The edge of change'.

Having taken care to distinguish between ongoing happening on the one hand, and change on the other, Schatzki writes about how practices, and bundles of practice, shape and in a sense generate each other. In his account, relevant processes include those of coalescence (in which rules, norms, understandings and arrangements combine), hybridisation (in which practices merge) and bifurcation (in which they split apart). In all of this, the complex relation between stability (which itself requires ongoing reproduction) and more transformative forms of evolution is very much in view, as are differential, but parallel, rates and types of change. These opening chapters elaborate concepts and definitions of social practice and frame much of what comes next, including more detailed empirical studies of transition and persistence in the materials and resource intensities of what people do.

The materials of practice

The goal of 'dematerialising' daily life, for example, by a factor of ten,[2] has been around for a while. The basic idea is that of developing technologies and techniques that allow people to do the same or more, but with less environmental impact. In exploring the uses of information and communication technologies (ICTs), Inge Røpke and Toke H. Christensen's chapter on 'Transitions in the wrong direction: digital technologies and daily life', and Sari Yli-Kauhaluoma *et al.*'s chapter on 'Mundane materials at work: paper in practice', work as a pair. Both follow the adoption of new technologies, showing how novel material elements do, and do not, transform existing practices. The examples and cases discussed illustrate the resource implications of some of the more abstract processes of linkage, hybridisation and bifurcation introduced by Warde and Schatzki. In addition, these two chapters work with the suggestion that social practices depend on the active integration of elements, including meanings, competences and materials (Reckwitz 2002; Shove *et al.* 2012). In this context, the common challenge is to show how ICTs become embedded and how this affects related patterns of energy and resource consumption.

Two rather different pictures emerge. The first is one in which ICTs are quickly integrated into practices as diverse as horse riding, bird watching and keeping in touch with friends and family. As described, widespread use of ICTs tends to soften time-space constraints, leading to an intensification of daily life. This is not the only possible outcome. Røpke and Christensen argue that in other social and economic circumstances, ICTs might transform practices in very different ways, and in ways that do reduce consumption. The present trend is, nonetheless, one in which more is packed into the day, and more energy is used as a result. By contrast, the second narrative, based on a small scale study of office life, emphasises the persistent importance of paper as a medium through which separate practices are coordinated and organised. Yli-Kauhaluoma *et al.* contend that detailed analysis of these cross-cutting functions helps explain why the paperless office remains a myth, and why ICTs have *not* led to the dematerialisation of working life. Røpke and Christensen describe processes of rapid appropriation, whereas Yli-Kauhaluoma *et al.* emphasise much slower forms of co-evolution. However, both demonstrate the subtlety of material innovation in practice.

Though not spelled out in quite these terms, both chapters question simple, essentially technological, goals of dematerialisation. They do so by reminding us that ICTs (and paper) have multiple roles, figuring as elements necessary for the conduct of specific practices, but also bridging between different practices, and mattering for when and where these are reproduced. This argues for a correspondingly subtle analysis of the relation between materiality (including resources and forms of consumption that are important for carbon emissions and ecological footprints) and a multiplicity of practices. Later chapters take this discussion further, but before turning to these more systemic analyses, the next two chapters concentrate on how people become practitioners and hence how practices spread.

Sharing and circulation

Social practices such as bathing, showering, driving and cycling vary in terms of how many people reproduce them, and where they are routinely enacted. In crude terms, we might say that some practices are bigger, more extensive, and more widely and frequently performed than others. To give two simple examples, at present in the UK, showering is more common than bathing, and driving is more common than cycling (in terms of time spent and kilometres travelled). Moving beyond the UK, what Urry (2004) refers to as the 'system of automobility' has become established across most countries of the world, so much so that driving has, in a sense, overtaken practices of cycling and walking. In theory, we might characterise the present state of any one practice at a particular point in time by mapping its geographic spread, the rate of recurrent performance, and the sheer number of 'carriers' or practitioners involved (Reckwitz, 2002). Should such maps exist, and should they be regularly revised, they might be used to track the movement and global circulation of more and less resource intensive practices (Shove 2009). Although there is no such atlas, understanding how practices spread remains important in thinking about where, and how, lower carbon ways of life might take hold.

Allison Hui's empirical study of how ashtanga yoga and leisure walking have developed to become globally popular pursuits provides some clues. Hui suggests that the spread of a practice, its 'circulation', is related to the separate movements of people (carriers), and of materials and knowledge, these being 'elements' of practice, as defined by Reckwitz (2002) and by Shove *et al.* (2012). Hui's central claim is that the circulation and persistence of requisite elements is a necessary condition for the diffusion of practices across space and time. A second related observation is that much travel is undertaken as a necessary part of doing certain practices. At first sight, neither yoga nor leisure walking require extensive resource consumption, but for some, being a practitioner prompts and justifies long and frequent journeys. As Hui explains, patterns of travel (and the destinations involved) are bound up with the historical development and circulation of the practices in question. For example, the history of leisure walking is such that the practice is closely associated with quite specific locations (hills, countryside etc.). Likewise, the manner in which yoga is organised and taught (by gurus and through first-hand experience) means that devoted practitioners are obliged to travel if they are to perfect their technique in the proper manner. In both cases, the history of the practice's development and diffusion is of some significance for where and by whom it is enacted, and hence for the forms of mobility that are subsequently required.

Russell Hitchings, who is also interested in how practices circulate, takes a different approach, focusing on the forms of learning and sharing that occur through 'communities of practice' (Lave and Wenger 1991; Wenger 1998). Lave and Wenger write about how novices become fully-fledged practitioners, and about how practices develop as cohorts of newcomers and old hands interact. Like others who discuss the careers of people and practices, these authors emphasise the socially shared character of seemingly personal experience and expertise. In Chapter 6, Hitchings revisits two recent research projects with these ideas in mind. The first study, which focused on how office workers keep cool in summer, concluded that shared practices were an outcome of collective learning, much along the lines that Lave and Wenger describe. By contrast, the second project, which was about how older people keep warm in winter, revealed a wider range of rather more private arrangements. Comparison of these two cases leads Hitchings to conclude that existing social networks are crucial for how strategies and practices of comfort circulate and change. As it happens, the conventions of comfort that office workers share are increasingly dependent on air conditioning: in this case the community is one that fosters and reproduces energy intensive practices. By contrast, older people, who might benefit from sharing skills in low-energy methods of keeping warm, often kept their habits private; they were not part of any comparable community of practice.

This discussion highlights the ambivalence of 'community' as a site of ecological innovation. It suggests that some communities, like those to which office workers belong, constitute networks through which *more* resource intensive practices circulate, whereas others may preserve pockets of tradition and variety. This might be 'good' in social and environmental terms, but it might also be problematic. Either way, the project of bringing concepts of community and practice together draws attention to the dual significance of social networks as conduits through which practices travel, and as outcomes of past patterns of recruitment. This complicates accounts in which 'communities', usually defined in geographical terms, figure as crucibles of pro-environmental change.

Relations between practices

The next three chapters mark a change of gear. Rather than focusing on how practices develop and circulate, one at a time, they concentrate on relations between practices: on how systemic transformations come about, on the work involved in making new practices normal within such systems, and on the extensive 'geometries' of practice that characterise daily life in contemporary industrialised societies.

Matt Watson makes a start on this agenda by speculating about what it might take for cycling to displace driving on a societal scale. The methodological strategy of taking this seemingly fanciful idea seriously forces him to take stock of the changing relationship between systems of velo- and automobility over time. Rather than representing systemic transition as a single evolutionary process, as is usually the case (Geels 2002), the relation between co-existing systems takes centre-stage. In taking this approach, Watson shows how theories of social practice and of sociotechnical innovation might be combined.

His main argument is that innovations in practice depend on patterns of recruitment and defection (for example recruitment to driving, defection from cycling), and that sociotechnical transformations are, in the end, underpinned by such processes. Although competitive relations between practices are crucial, it is clear that driving does not simply take the place of cycling. As Watson goes on to show, meanings of 'fast', 'convenient' and 'slow' modes of transport are outcomes of the changing *relation* between cycling and driving, and between the practices and infrastructures of which past and present systems of velo- and automobility are composed. From this it follows that opportunities for deliberate policy intervention (for example, to promote cycling) are themselves an emergent product of ongoing relations and dynamic processes within and between these co-existing systems. To some extent this is obvious: strategies that make sense when 40 per cent of local trips are by bike, are evidently not the same as those that might be adopted when this figure has fallen to one or two per cent. A less obvious conclusion, but one that comes from this discussion, is that policy making, past and present, does not stand outside the systems of practice that it seeks to influence, but is instead integral to them. In other words, policy interventions have effect when integrated into ongoing, already dynamic processes of recruitment and defection that are in turn shaped by forms of path dependence, lock in and feedback.

Concluding that systemic transitions have somewhat emergent lives of their own does not prevent us from recognising and assessing the normative character and practical significance of deliberate attempts to steer practices in one direction or another (a theme explored in Chapter 10). As represented here, the project of understanding how people become carriers of practice (and how practices compete with each other) provides a means of understanding how power is exercised and how inequalities are reproduced. This occurs at many levels and, as detailed in the next chapter, there is no escaping the politics of technology and no denying the commitments and judgements that are, of necessity, folded into the dynamics of social practice.

Julien McHardy's chapter is about the work involved in bringing electric cycling into being, and establishing it as a specific type of practice. In the European context, electric bikes are promising technologies (van Lente 1993; Brown, *et al.* 2000): though not yet 'normal', they promise to extend the range and scope of cycling and perhaps reduce reliance on the car. Electric bicycles are not viewed as the harbingers of a radically new regime of velomobility, but they do figure in future oriented discourses of sustainability.[3] At the same time, the question remains, what *is* electric cycling and what might it become? In order to address this question, McHardy zooms in on the technicalities of testing and comparing competing models. As he explains, the process of testing electric bikes simultaneously anticipates and constructs electric cycling, defining the riding styles of imagined cohorts of future cyclists and ensuring that those who participate in the test faithfully reproduce this not-yet actual practice. The test process depends on managing and containing variety in order to define a somewhat stable form, this being a necessary step if performances are to be compared and if electric cycling is to be presented as a recognisable entity that exists across time and space. There is nothing special about this particular test procedure (many tests work this way), but there is something intriguing about how relations between actual and future cyclists and bicycles are negotiated in the process. These relations, along with also imagined interactions between electric cycling and driving, are inscribed in test protocols that delimit the contours of electric cycling as a practice and that reveal/establish how this practice is/is not to be normalised and positioned within and alongside existing systems of urban and semi-urban mobility.

We do not know what will happen beyond the confines of the test, or how electric cycling will fare in the longer run. Whatever transpires, electric cycling will be shaped by the standardising work undertaken by the testers' protocols and methods and the impact these have on product design, and on the terms in which electric cycling is framed for future users. Though not the only possible configuration, the result is a package in which actual bikes and imagined riders are bundled together and constituted as an increasingly stable configuration known as 'electric cycling'.

Grégoire Wallenborn's more philosophical chapter takes McHardy's discussion of these human-non-human hybrids to heart. It is one thing to suggest that practices constitute the basic unit of social enquiry, but who, or what, then constitutes 'the practitioner'? Is electric cycling enacted by the rider alone, or by an always fluid combination of rider and bicycle? (sometimes one does the work, sometimes the other). This line of thinking opens the way for a more detailed analysis of what Wallenborn refers to as the 'extended body', this being the human body, plus everything else that is implied in the performance of a practice. Wallenborn distinguishes between five generic 'places' in which a body might be extended and concludes that the 'geometry' of a practice is defined by the specific intertwining of these 'topographies'. His five-fold scheme provides an analytic framework with which to represent the changing outlines both of practices, and of the extended bodies that enact them. It also provides a set of terms with which to contemplate future possible scenarios, including ones in which practices and bodies are systematically less extended, and in which qualities of life are instead defined with reference to the richness and variety of actual and potential 'assemblages'. Wallenborn's contribution takes us away from particular practices to forms of containment, circulation and flow that characterise systems and complexes of practice. In so doing, it brings questions about the relations between extension (including delegation to other people and things), resource intensity and well being into the frame.

The last two chapters in the collection take these topics as their central concern. Andrew Sayer's contribution outlines a number of challenges facing those who are interested in concepts of sustainable practice. These have to do with the role and status of values, with how capitalist systems of profit making fit into the picture, and with how concepts of well being and social practice might connect. In responding to some of these challenges, Gordon Walker explores the potential for combining Sen's work on capability and justice with the literature on social practice. In working through these ideas, both authors discuss the limits of social theories of practice as currently formulated, and both consider ways in which some of these might be overcome.

Sustainability, inequality and power

Serious consideration of political economy and power, and questions of morality and justice have not figured prominently within the 'practice turn'. However, any systemic transition towards sustainability will take place in the context of present patterns of inequality and power. In addition, significant change of the scale required is itself sure to generate 'winners' and 'losers' (Shove and Walker 2010). There is no doubt that these are issues with which a discussion of sustainable *practice* needs to engage, but how are these to be approached and framed? For example, are values and judgements best understood as being integral to practice and as outcomes of practical experience as in MacIntyre's view (1985), or as commitments that stand somewhere outside the realm of practice? This ongoing discussion draws attention to the necessarily political and contested character of efforts to steer or modify the trajectories of social practice, many of which are central to the functioning of capitalist societies. It is in this context that Sayer discusses the place of individuals' reasons and values. He is particularly concerned that the representation of people as the (mere) carriers of practice risks positioning them as passive hosts, rather than as evaluative, judgemental and creative beings. Though at one extreme this might be the case, carriers also figure as essential sources of variation, transformation, conflict, contest and change. The question is therefore not whether people-as-the-carriers-of-practice have values or make moral judgements (they clearly do), but rather, how these are to be conceptualised.

In the second part of his chapter, Sayer asks whether theories of practice can address questions of economic and corporate power. It is fair to say that some of the recent literature is preoccupied with the details of consumer behaviour and with the immediacies of doing – for example, doing heating, doing waste management or whatever (Gram-Hanssen 2010; Hargreaves 2011). In writing of this kind, questions about the provision and supply of necessary elements of practice – including sources and forms of energy and what Wallenborn describes as the globally 'extended bodies' involved – disappear from view. As Sayer points out, it is easy to limit an analysis of 'practice' to a close study of situated performance, whether of cooking and heating, or of financial management. And it is true that when concepts of practice are cut short in this way, it is difficult to see how they might help in understanding evidently relevant trends in business, profit making and politics.

It is in this context that Sayer makes a case for bringing political economy back into a discussion of practice. He does not go into detail about what this might involve, but if the positions developed by other contributors to this collection are anything to go by, responses are likely to vary rather widely. For example, some might argue that problems of macro-economic transformation are not problems that theories of practice can usefully address. Others might distinguish between practices, on the one hand, and social and economic contexts/forces on the other, perhaps going on to detail the relation between the two. Alternatively, there may be no need to bring political economic concerns 'back' on the grounds that they have never been absent. From this point of view, processes of accumulation and conflict and patterns of inequality are outcomes of, and are central to, the lives of practices. As such, they are crucial in understanding how practices circulate and become dominant, and how complex systems of practice develop and endure. These various interpretations reflect underlying differences in how practices are conceptualised. There is no one theory of practice, but as demonstrated in other parts of this book, some versions are capable of explicating unsustainability and injustice at this macro scale.

This potential needs to be developed and, in the final chapter, Gordon Walker gives a sense of what this might involve. Sayer and Wallenborn both conclude by suggesting that a more sustainable future depends on identifying and promoting 'frugal' practices that enhance and contribute to well being. Walker homes in on the issue of what constitutes a good and worthwhile life, doing so in a manner that allows him to explore connections between social theories of justice (in particular, Sen's capabilities framework) and of practice. In working these ideas through in greater detail, Walker makes use of the distinction between practice-as-entity (that is, as something that is recognised and that persists across space and time), and practice-as-performance (moments in which practices-as-entities are enacted), revisiting Sen's notions of functioning and capability with reference to these ideas. As with other chapters, the results of this exploratory exercise are promising and generative. Representing social practices as the doings through which good and worthwhile lives are achieved, seems to provide a means of approaching the normative and ethical questions that surround deliberate efforts to promote sustainability.

These final chapters point to an emerging debate about the limits and potential of practice theory. This revolves around two positions. The first suggests that issues of political economy and power have been bypassed because theories of practice are ill-equipped to deal with them. The second suggests that concepts of practice can be developed in this direction, as Walker's chapter demonstrates.

Sustainable practice

Some of the first efforts to represent and analyse sustainable *practice* were developed in response and reaction to dominant paradigms emphasising attitude, behaviour and consumer choice (Shove 2010). The chapters in this book show how much more there is to do, and how much potential remains. As we have begun to see, theories of practice (broadly defined) are capable of feeding into questions of *sustainability* from a number of different angles: contributing to understandings of justice, transition, innovation and political economy, as well as consumption. This is relatively new territory, but the chapters in this book provide terms and concepts that are useful in thinking about how certain practices dominate and how complexes and systems of practice become embedded and established at a societal scale. Further work is required to describe and characterise forms of accumulation (for example, of money, competence, infrastructure, etc.), and of distribution and access that follow from past patterns of practice and shape future possibilities. This depends on understanding (not overlooking) the lives of people who carry, contest, negotiate and manage multiple practices at once, and on understanding how values and meanings of well being are embedded in processes of recruitment, defection and ongoing transformation. In dealing with aspects of this agenda, the chapters included in this book give a sense of what a more comprehensive account of *sustainable practice* might entail. This is unfinished business, but in putting this collection together we have made a start.

Notes

1 Available online at: www.ecologicalfootprint.com/.
2 Available online at: www.factor10-institute.org/.
3 Available online at:www.sustainable-mobility.org/.

References

Brown, N., Rappert, B. and Webster, A. (eds) (2000) *Contested Futures: A Sociology of Prospective Techno-Science*, Aldershot: Ashgate.

Cohen, M. and Murphy, J. (eds) (2001) *Exploring Sustainable Consumption*, Amsterdam: Pergamon Press.

Geels, F. W. (2002) 'Understanding the dynamics of technological transitions: a co-evolutionary and socio-technical analysis', Enschede: Twente University Press.

Giddens, A. (1984) *The Constitution of Society*, Cambridge: Polity Press.

Gram-Hanssen, K. (2010) 'Standby consumption in households analysed with a practice theory approach', *Journal of Industrial Ecology*, 14(1): 150–165.

Gronow, J. and Warde, A. (2001) *Ordinary Consumption*, New York: Routledge.

Hargreaves, T. (2011) 'Practice-ing behaviour change: Applying social practice theory to pro-environmental behaviour change', *Journal of Consumer Culture*, 11(1): 79–99.

Lave, J. and Wenger, E. (1991) *Situated Learning: legitimate peripheral participation*, Cambridge: Cambridge University Press.

MacIntyre, A. (1985) *After Virtue*, London: Duckworth.

Reckwitz, A. (2002) 'Toward a theory of social practices: a development in culturalist theorizing', *European Journal of Social Theory*, 5(2): 243–263.

Schatzki, T., Knorr Cetina, K. and von Savigny, E. (eds) (2001) *The Practice Turn in Contemporary Theory*, London: Routledge.

Shove, E. (2003) *Comfort, Cleanliness and Convenience: the Social Organization of Normality*, Oxford: Berg.

Shove, E. (2009) 'Everyday practice and the production and consumption of time', in E. Shove, F. Trentmann and R. Wilk (eds) *Time, Consumption and Everyday Life: Practice, Materiality and Culture*, Oxford: Berg.

Shove, E. (2010) 'Beyond the ABC: climate change policy and theories of social change', *Environment and Planning A*, 42(6): 1273–1285.

Shove, E. and Walker, G. (2010) 'Governing transitions in the sustainability of everyday life', *Research Policy* 39, (4): 471–476.

Shove, E., Pantzar, M and Watson, M. (2012) *The Dynamics of Social Practice: Everyday Life and How it Changes*, London: Sage.

Southerton, D., Chappells, H. and van Vliet, V. (2004) *Sustainable Consumption: the Implications of Changing Infrastructures of Provision*, Cheltenham: Edward Elgar.

Spaargaren, G. (2011) 'Theories of practices: agency, technology, and culture: Exploring the relevance of practice theories for the governance of sustainable consumption practices in the new world-order', *Global Environmental Change*, 21(3): 813–822.

Urry, J. (2004). 'The "System" of Automobility', *Theory, Culture & Society*, 21(4–5): 25–39.

van Lente, H. (1993) 'Promising technology. The dynamics of expectations in technological developments', University of Twente, PhD.

van Vliet, B., Chappells, H. and Shove, E. (2005) *Infrastructures of Consumption: Environmental Innovation in the Utility Industries*, London: Earthscan.

Voß, J. and Bornemann, B. (2011) 'The politics of reflexive governance: challenges for designing adaptive management and transition management', *Ecology and Society* 16(2): 9.

Warde, A. (2005) 'Consumption and theories of practice', *Journal of Consumer Culture* 5(2): 131–153.

Wenger, E. (1998) *Communities of Practice*, Cambridge: Cambridge University Press.

Part I
How are practices defined and how do they change?

2 What sort of a practice is eating?

Alan Warde

Introduction

The close of the 20th century witnessed a renewal of interest in theories of practice, an interest sufficient to make possible the rhetorical announcement of the birth of *The Practice Turn in Contemporary Theory* (Schatzki *et al.* 2001). The move has had some considerable significance, even if Omar Lizardo (2009: 714) may have overstated the case when proclaiming that 'It can be said without much danger of exaggeration that practices now play as central a role in sociological thinking as values and normative patterns did during the functionalist period.'

Some foundations had been laid three decades earlier. In a widely cited article about trends in theory in anthropology, published in 1984, Sherry Ortner (1984: 127) observed that 'a new key symbol of theoretical orientation is emerging, which may be labelled "practice"' (or 'action' or 'praxis'). She described its emergence from the intersection of theoretical schools in anthropology formed in the 1960s, and inter-disciplinary Marxism and political economy in the 1970s. The principal authors credited with the development were two sociologists, Pierre Bourdieu and Anthony Giddens, the anthropologist, Marshall Sahlins and the social theorist, Michel Foucault. A primary common objective was to account for action in a manner that was complementary to the study of systems and structures (Ortner 1984: 147–8). In European sociology, the legacy of structuralism figured strongly in the intellectual context in which the problem of conceptualizing the relation between structure and agency was given proirity. The concept of *Praxis* played a central role, and was conceived as a bridging device between equally flawed holist and individualist explanations. Nevertheless, holism was more comprehensively attacked than was individualism; discomfiture with a previously rampant structuralism always threatened that agency might be allowed to dissolve into personal autonomy, a temptation to which many succumbed. In the process, despite the positions adumbrated in works like *Outline of a Theory of Practice* (Bourdieu 1972) and *The Constitution of Society* (Giddens 1984) being seminal for social theory and inspiring empirical studies, the orientation towards the analysis of practice per se palled. It was left to a second generation of advocates to re-centre the concept of practices.

The most distinctive feature of the Second Coming of practice theory, most clearly articulated by Theodore Schatzki (1996, 2002), is its injunction to view practices as the fundamental unit of social scientific analysis.[1] Preferred to individual action – which has dominated economics, psychology and most of micro-sociology in neo-liberal times – practices are proposed as the central scientific object of study and as a means to avoid a collapse into methodological or ontological individualism. (A similar problem simultaneously inspired other innovative sociological approaches; for example, McFall *et al.* (2008) develop a culturalist interpretation of Weber and the works of Foucault and Elias; DeLanda (2006) approaches the issue by way of 'assemblages'; and Fine (2010) via situated action).[2] However, how best to examine practices remains controversial. Agreement among advocates of the practice approach is limited (Schatzki *et al.* 2001:2, 2011), although in general all give precedence to practical activity as the means by which people secure their passages through the world thereby emphasising doing over thinking, practical competence over strategic reasoning, mutual intelligibility over personal motivation and body over mind.

In this paper, I want to explore a few of the implications of taking practices as the basic unit of sociological analysis. I am primarily concerned with the methodological issue of how to identify a practice which is to become the object of study, particularly its boundaries. To do so, however, raises ontological issues concerning the substantive characteristics of a practice. In addressing these issues, I draw to some degree on an earlier paper which provided a brief sketch of what difference it might make to the analysis of consumption if a practice-theoretical approach were adopted (Warde 2005). In this earlier article, my initial intention had been to present a schematic account (through a focus on the practice of eating) of the application of second-generation concepts, upon which topic I had conducted extensive empirical research in Britain. However, this proved to be insurmountably difficult, so instead I used the example of motoring, which provided a satisfactory means of illustrating the analytic potential of the theory. The current chapter is largely inspired by a wish to understand why eating proved to be such an intransigent case. The problem, I argue, is that it is a particularly complex type of practice, which I will call a *compound* practice. To develop the case, I return initially to my characterization of the practice of motoring, where I found it helpful to frame my account in terms of two distinctions to be found in the work of Schatzki (1996) and Reckwitz (2002): namely between *praktiken* (practices as entities) and performances, and between integrative and dispersed practices.

The nature of practice

Schatzki (1996) identifies two core features and two generic types of practice. First, he distinguishes practice as an organized nexus and practice as performance.

The first notion is of:

> practice as a temporally unfolding and spatially dispersed nexus of doings and sayings. Examples are cooking practices, voting practices, industrial practices, recreational practices, and correctional practices. To say that the doings and sayings forming a practice constitute a nexus is to say that they are linked in certain ways. Three major avenues of linkage are involved: (1) through understandings, for example, of what to say and do; (2) through explicit rules, principles, precepts and instructions; and (3) through what I will call 'teleoaffective' structures embracing ends, projects, tasks, purposes, beliefs, emotions and moods.
>
> (Schatzki 1996: 89)

Important to note here is that practices consist of both doing and sayings, suggesting that analysis must be concerned with both practical activity and its representations. Moreover we are given a helpful depiction of the components which form a 'nexus', the means through which doings and sayings hang together and can be said to be coordinated.

The second sense, practice as performance, refers to the carrying out of practices, the performing of the doings and sayings which 'actualizes and sustains practices in the sense of nexuses' (Schatzki 1996: 90). The reproduction of the nexus requires regular enactment. As Reckwitz describes it:

> a practice represents a pattern which can be filled out by a multitude of single and often unique actions reproducing the practice.... The single individual – as a bodily and mental agent – then acts as the "carrier" (Trager) of a practice – and, in fact, of many different practices which need not be coordinated with one another. Thus, she or he is not only a carrier of patterns of bodily behaviour, but also of certain routinized ways of understanding, knowing how and desiring. These conventionalized "mental" activities of understanding, knowing how and desiring are necessary elements and qualities of a practice in which the single individual participates, not qualities of the individual.
>
> (2002: 249–50)

Reckwitz thus appears to offer a stronger interpretation of the organized nexus, implying that practices are in some way coordinated entities, since they require individual carriers to instigate the performances necessary for their existence. Performances presuppose a practice. As Giddens put it:

> Human social activities ... are recursive. That is to say, they are not brought into being by social actors but continually recreated by them via the very means whereby they express themselves *as* actors.
>
> (1984: 2)

Schatzki also draws a distinction between dispersed practices and integrative practices. 'Dispersed practices' (1996: 91–2) appear in many sectors of social life, examples being describing, following rules, explaining and imagining. Their performance primarily requires understanding; an explanation, for instance, entails understanding of how to carry out an appropriate act of 'explaining', an ability to identify explaining when doing it oneself or when someone else does it, and an ability to prompt or respond to an explanation. This is about 'knowing how to' do something, a capacity which presupposes a shared and collective practice involving performance in appropriate contexts and mastery of common understandings, which are the grounds for a particular act being recognizable as explaining.

'Integrative practices' are 'the more complex practices found in and constitutive of particular domains of social life' (1996: 98). Examples include farming practices, cooking practices and business practices. These include, sometimes in specialized forms, dispersed practices, which are part of the components of saying and doing and which allow the understanding of, say, cooking practice, along with the ability to follow the rules governing the practice and its particular 'teleo-affective structure'. These are examples which, Schatzki suggests, will generally be of most interest to sociologists.

Schatzki (1996: 98–110) offers several criteria for recognizing the existence of integrative practices.[3] For present purposes three claims are especially germane. The first is that performances can be read as correct and acceptable, even when innovative (1996: 101–2). The second is the proposition that 'practices are social by virtue of exhibiting coexistence with indefinitely many other people' (1996: 105). The third is that some entity exists which is not simply in the minds of individuals: 'the organization of an integrative practice is out there in performances themselves not in the minds of actors' (1996: 105). The first two apply to both dispersed and integrative practices, the last only to integrative practices.

Some close attention needs to be paid to what it means for a practice to be 'organized'. For Schatzki, in the first instance, organization refers to the establishment of a 'nexus of doing and saying', comprising understanding, rules and teleo-affective structure, which can be seen to fit together when observed in competent performances. Organization seems primarily to be a matter of shared prescriptions for the coordination of specific performances. The performances are instances which confirm that 'the organization of an integrative practice is out there in performances themselves'. However, it is not entirely clear what it is 'out there' that exhibits the practice's 'coexistence with indefinitely many other people' and thereby allows sharing. Clarification, it seems to me, might come from subjecting the notion of organization to sociological analysis and examining how practices are socially coordinated.

Schatzki is very hesitant about the matter, but I side with Reckwitz in considering it as essential to a sociological version of the theory that we think of practices as *entities*. The distinction between a practice and its performances is especially important because every performance of 'X' is singular and particular, yet it is essential to be able to determine whether any such given performance truly belongs to the category of 'X-ing'. That is not straight-forward, and scholars carrying out empirical investigations frequently find drawing the boundaries around 'X-ing' highly problematic.[4]

Performances are not inherently problematic phenomena for sociology. They are stock-in-trade objects of investigation for micro-sociologies of several different types. Practice theories say that they are necessary for the reproduction and development of a practice, that they are things that people and other agents do, that they are not done in the same way by everyone – they are in many instances individual, innovative, improvised, as well as being context-dependent. One felicitous characterization might be to say that while the majority of performances are improvisations within a practice, they are mostly still recognizable as unambiguous examples of that particular practice. More challenging, is the provision of a convincing account of the ontological and epistemological status of the practice that subtends the performances. Some of the things we might want to know include how practices should be characterized in terms of their elements and their arrangements, and also how they should be studied. This, I believe, requires explicit acknowledgement of the fact that practices are established, delimited, reproduced and organized through *social* processes of practical coordination. That, it seems to me, is fundamental to analysis of the nature of practices (*praktiken*).

Eating as a problematic case

Since I would like to develop a theory of eating, it is important to me to know whether it is helpful to understand eating as a practice. If so, is it an integrative or a dispersed practice, and if not how should it be characterized? At first glance eating is a physiological activity associated with ingestion, and in some languages this is its primary meaning. However, it is almost impossible to detach that facet from issues of what is eaten in which social circumstances. Consequently dictionaries describe it as both a physical and a mundane social activity. This may perhaps be a source of my own difficulty in accommodating it with practice theory.

First, maybe eating is too simple an activity to be amenable to analysis in practice theoretical terms. If equated with ingestion, it might be maintained either that it is not learned (since it is instinctive), or not learned consciously. However, eating is otherwise socially complex; it is an elaborate domain of everyday life, an activity not understandable without reference to its social embeddedness. It is hard to give an account of eating without referring to the selecting of foodstuffs, preparing dishes, making social arrangements for meals and aesthetic judgments about taste. Indeed, it is clear that even bodily techniques of food incorporation are culturally variable (e.g. Bourdieu 1984; Elias 1969; Tuan 2005). Coming to grips with that complexity suggests that the problem is not its simplicity.

A second possibility is to consider eating as a dispersed practice. The basic difference between a dispersed and an integrative practice is clear enough at the extremes. A dispersed practice operates in association with, and often through, other integrative practices. To recall, Schatzki's (1996: 91) examples of dispersed practices include 'describing, ordering, following rules, explaining, questioning, reporting, examining, and imagining'.[5] A dispersed practice finds its particularity or concreteness when incorporated into others. Most integrative practices will depend upon a great many other dispersed practices (e.g. explaining, describing, reporting, etc.). If explaining and reporting are dispersed practices, they find their particularity as, for instance, sociological explanation or research reporting. An integrative practice is more concrete and more substantial in the sense that it is constituted solely, or even primarily, by shared understandings. Examples of integrative practices are 'farming practices, cooking practices and business practices' (Schatzki 1996: 91). The implication is that their teleo-affective properties (especially shared purposes of engagement in the practice) are definitive and, in the examples offered, the activities involved are most often task-oriented, instrumental activities, which are readily subsumed under the framework. The central example in Schatzki's (2002) demonstration of the applicability of his ontological theory is how the Shaker community organized the production and distribution of herbal medicines. As an integrative practice it was complex, with each performance presupposing competence in several others besides the focal activity of distilling and bottling medicinal compounds. Nevertheless, the purpose or objective of performances was transparent, and success or failure was easy to determine with reference to chemical composition, purity and therapeutic efficacy – not to mention profitability.

Could eating be treated as a dispersed practice, as one which takes concrete form only in relation to other practices, such that eating is a part of medicine, a part of cooking and a part of socializing? It is certainly possible to learn about aspects of healthy eating or good taste indirectly from the disciplines and crafts of medicine, nutrition, etiquette and cookery. However, eating does not appear to be like other typical examples of a dispersed practice (such as describing, questioning or imagining), for it seems to have a bounded and substantial presence which, for instance, permits it to be recorded in time-use studies; one can unambiguously allocate episodes of eating to a period in the day, but one cannot readily do the same for episodes of questioning or imagining. Must eating then be, by default, an integrative practice? Eating as an activity has many of the characteristics of an ideal typical integrated practice: it is a major activity of daily life; it is performed very often; it involves doing and saying; it has a large vocabulary devoted to it; much is written about it; it is routine and habitual in several senses; it involves understandings, procedures, specialised equipment and purposes; and it is instantly recognizable when encountered. However, my doubts arise from making comparisons with both herb manufacture and motoring, for it would seem that eating is neither coordinated nor regulated in the same fashion. Most immediately, in Britain at present there exists no shared understanding of what it means to eat well. Something is amiss regarding the normative structuring of the practice, its standards and its subtending motivations. For while there is widespread practical understanding of what it means to eat, there are no shared *standards* governing the activity. Clarification, it seems to me, might come from subjecting the notion of organization to sociological analysis.

Integrative practices as socially coordinated entities

Social organization is a necessary condition of the existence of a practice; one which is accomplished by interested parties. One appeal of considering practices as entities is that processes involved in the social coordination required to steer and regulate performances can be brought to light. Performances become practices through explicit engagements and commitments to the normative specifications of appropriate performances. This is the nub of Schatzki's crucial and justifiable claim that performances of an integrative practice can always be subjected to judgements of correctness and acceptability. The implication is that standards are publically recognizable. One source of public recognition is the fact that criteria of acceptability are very often expressed in, and are core to, processes of explicit formalization and codification. Codification is a matter of specifying the objectives or purposes in view in a domain of activity, and the ways to go about attaining such goals. Nowadays, this is usually achieved through documentation of rules, procedures and standards. Codification of an integrative practice like farming or cooking typically involves many different institutional processes. Often formalization and regulation is the result of the activity of formal organizations. Associations (for example professions, governing bodies of sports, statutory regulatory state agencies, not to mention educational institutions, etc.) play central roles in coordination. They prescribe rules, prohibit or discourage particular forms of behaviour, teach acceptable conduct and present prizes for excellence etc. While these are formal and often authoritative agents and not all integrative practices are constituted or steered by such associations, they are an essential component of the ordering, or organization, of many practices. Indeed, one of the ways in which we might recognise a practice is through the existence of formalized processes, procedures or artefacts directed towards specifying adequate performances.

One common means of formalization, which may facilitate improvement of individual performances in the light of the standards of a practice, is to describe, record and release for public circulation, instructions about how to do something, how to do it better and how to do it well. I have in mind artefacts like rule books, teach yourself primers, instruction manuals for improving performances, guide books, etc. These are sociologically interesting phenomena. Practice manuals give us robust prima facie evidence of the existence of a practice. For practices only exist where performances can attain to some standard of excellence and thus, for most individuals, be improved. Manuals also seem to provide solid evidence of the facticity (and common recognition) of the existence of some underlying foundations to correct or acceptable performances of a complex and widely shared practice.

This suggests one operational criterion for deciding whether we are confronting an integrative practice. Though, on the surface perhaps, a trivial hypothesis, I submit that to pass as an integrative practice it should be possible (in principle) for an activity to be included in the 'Teach Yourself' series![6] Teach Yourself books have a number of defining characteristics:

- They usually offer a simple or preliminary account of content or relevant know-how, presented in terms of rules or facts, in a manner suitable for novices to fashion and improve performance.[7] Note that the possibility of improvement indicates that performances can be changed and this implies that changing performances may contribute not only to the reproduction but also to the evolution of the practice.
- They outline the nature of and the means to acquire competence to deliver performances which would be recognized as adequate when relayed to a competent audience.
- They present the activity as *praktik*, that is, as a coordinated entity, with shared norms of performance.

The paradox of formalization

Exactly what it is that is 'out there' is critical to explaining how (some) practices are coordinated. It is interesting that this is not often observed in accounts which apply a theory of practice to micro-sociological analysis, perhaps because such a rule of thumb may initially appear paradoxical. The paradox of the process of codification is, of course, that competent practitioners do not perform 'by the book', by following rules, or by reference to a manual. Yet codification and formalization of a practice is normal, even routine. Collins (2010) astutely observes that the reason why tacit knowledge has attracted so much attention in recent years is that since the onset of modernity in Europe, societies have, through countless exercises in rationalization, attempted to make all knowledge explicit. Critically, though, this does not entail that performances will be uniform or correct imitations. The components ('out there') of the practice are evidence for particular activities having been developed into practices, but it cannot be assumed that manuals are much used in order to make decisions or plan courses of action or intervention. Codification of a practice may have relatively little *direct* effect on actual performances. That, however, does not render formalization irrelevant and, in the case of eating, it helps to pinpoint the nature of the practice.

Eating as a compound practice

Perhaps, therefore, a different formulation is required, which acknowledges that practices are shaped by the manner of their formalization; with consequences, especially, for the setting of standards of performance. Thus eating, except when viewed from the perspective of simple ingestion, might most plausibly be described not as part of another practice itself, but as drawing upon several integrative practices.[8] Eating, as Britons currently know it, presupposes the intersection of at least four integrative practices: the supplying of food, cooking, the organization of meal occasions and aesthetic judgments of taste. These are formalized in terms of nutrition, cooking, etiquette and gastronomy. Eating might therefore be seen as an especially complex practice. Conceived as an entity, one option would be to conceptualize it as emergent from a number of these other integrative practices. Even better, perhaps, would be to treat it as a composite or *compound* practice with multiple organizational underpinnings. Performances of eating are, in the latter view, a complex corollary of the intersection of four, relatively autonomous integrative practices.

The existence of four component practices probably makes eating difficult to coordinate at the level of both individual performance and collective institution. Consequently, eating performances have a tendency to be socially weakly organized, and even disorganized. Each of the component integrative practices has its own logic, and its own different coordinating agents. By comparison with motoring, eating is both weakly regulated and weakly coordinated. Regulations and organizations strongly define motoring, its standards, its justifications and its conditions for flourishing as a practice. Driving an automobile is founded upon an enforceable legal framework of acceptable behaviour. It is taught commercially in schools of motoring and competence is tested and licensed by the state. The infrastructure of road systems and traffic signs powerfully channels performance. Organizations of the state, in negotiation with the motoring lobby (automobile associations, guilds of truck drivers, etc.), produce a working consensus about acceptable parameters of the activity. Although there are different styles of driving, few motorists flout the norms of traffic management; norm and normality are in close alignment. Eating, by contrast, is more loosely framed; more a matter of convention than authoritative regulation, neither formally taught nor accredited, occurring mostly in private, not requiring constant and second-by-second coordination with strangers and not (despite social movements and consumer bodies) subject to direction and control by powerful organizations of practitioners or regulatory agencies. Although driving styles vary, there is a fair level of consensus about what it means to be a good driver. What it means to eat well is much less clear.

Arguably, the current disorganization of eating is historically specific. Its component integrative practices have evolved over many decades, even centuries, but have come together sufficiently for eating to be recognizable 'out there' in performances. In some circumstances, they come together to constitute an entity, which is socially coordinated in such a manner that an organized nexus can be identified. But at other times and in other places, coordination is weak. There are certainly historical and regional variations in the ways that these four integrative practices are put together. One instance of a high level of social coordination is the case of France, where until recently there were formal, institutional, intellectual and artefactual modes for specifying how these component activities should be organized and arranged in some authoritative manner; a combination of regional produce (*terroir*), French cuisine (cooking), the bourgeois family meal and an intellectual and sensual interest in eating and taste codified as gastronomy (Ferguson 2004; Trubek 2000; Warde 2009). By contrast, in the UK (now and in the past) eating is a weakly coordinated practice. For instance, food supply is mostly coordinated by supermarkets, with some government regulation, and some effects from contestation by social movements for safe and healthy food. Supermarkets make for a fissiparous domain, driven by selling not eating; major suppliers do not care what people eat, but rather aim only to provide customers with choices. Meals have, in the past, provided a strong temporal and social structuring to eating; but while etiquette, manners and rules about companionship and regular mealtimes may once have held sway, they are currently often flexible and informal. Moreover, although elements of a system for the judgment of taste may have emerged recently as a function particularly of the expansion of eating out (Warde 2009), the terms for comparison of objectives and standards, never prominent in Britain, are not yet widely established. Only cooking, despite having several agencies at work affecting its formalization, inter alia recipe books, educational institutions and the celebrity chef system, is relatively highly coordinated as a practice. Generally speaking, however, because these integrative practices have developed at different speeds and in relation to different logics, they are weakly coordinated.

One corollary of these observations might be to investigate the effectiveness and configuration of agencies specifically seeking to promote co-ordination. There are few agencies operating in the UK that might accomplish what the literate tradition of gastronomy achieved in France (Ferguson 2004). Perhaps the Slow Food Movement could be taken as a latter day attempt to frame the constituent components into a coherent compound practice of eating. It seeks to influence the supply of materials, cooking techniques, temporal rhythms and convivial eating, in light of an intellectual justification for reform of the eating habits associated with fast food and industrial farming (Petrini 2001). A more prosaic and mundane form of coordination is orchestrated by the catering trades, which create a partly integrated system of supply, preparation, social format and aesthetic standards. This is institutionalised through restaurant guides, celebrity chefs, the wholesale trade and innovation in relation to the combination and presentation of food. As a consequence, eating out tends to have greater coherence (see Warde 2004) than domestic practice.

In Schatzki's terms, eating is not an integrative practice, though it may become so in certain circumstances. The current structure of eating, and other similar practices, suggests that another concept is needed. To be an integrative practice, an activity requires appropriate social supports and sustenance, social coordination and organization. An established practice is an historical product, even an achievement (cf. MacIntyre 1985). From one persepecive, compound practices may be considered as failed, or not yet worked up, integrative practices. The several integrative practices that underpin eating are the effect of groups of practitioners usurping elements of what was once a more general field of activity and formalizing understanding, procedures and commitment to fashion a more specialized integrative practice. This leads towards a model of competition and contestation, and perhaps recommends the concept of social fields (Bourdieu and Wacquant 1992; Martin 2011). Certainly, capturing the dynamics of the process will be difficult if the institutionalized properties associated with the social regulation and coordination of practices are neglected.

Conclusions

Eating in the UK and, possibly increasingly across the Western world, is a weakly organized and weakly regulated practice. It is subject to the vagaries of the process of competition between several better focused and more tightly organized practices, and it is vulnerable to disorganization. This may be taken as grounds for coining a new concept to better describe complex and heteronomous practices like eating. Interventions to socially coordinate practices may gel poorly with the organized nexus made manifest in performances of that practice. In line with the fundamental axiom that performances and practices stand in a recursive relationship to one another, it may therefore be best to consider contemporary eating as an example of a compound practice. That is to say, the practice is recognizable through its performances, but judgment of their correctness (their conformity with standards) cannot be made with reference to a single integrative practice. The normative forces governing eating are divergent in their injunctions, even mutually contradictory. It seems likely that other major domains of everyday life, such as those identified in time diaries (travelling, recreation, working, civic participation, etc.) for instance, will demonstrate similar properties when subjected to analysis from a practice-theoretic perspective.

Irrespective of whether the concept of compound practice has merit, it is important to stress that the organization of practices, in Schatzki's sense, is itself the outcome of contestation between social groups over the right to define the standards and procedures which prevail in a given period. The complicating feature of eating is that there are several integrative practices that claim jurisdiction over eating performances. The doctrine of modern nutrition proposes different standards to those of aesthetic and lifestyle movements concerned with taste, and also to those of advocates of the valuable affective qualities of the family meal. Thus, in the example of Britain, particularly because it lacks a tradition of gastronomy, the absence of binding formal rules or handbooks teaching the art of eating has tended to result in weakly coordinated performances.

Second, a better understanding of the nature of the intersections between practices is essential. By addressing a rather troublesome example, I have (in this chapter) merely scratched the surface. Mutual intersection is a vital part of the explanation of the dynamics of practices. The existence of compound practices is a function of the dynamics of practice formation. Practices are not given, but are fabricated and consolidated in cultural contexts. Forces at work operate to turn compound practices into integrative ones and vice versa. We should consider more carefully how practices are socially and purposefully coordinated and what difference that may make to their mutability.

Third, formalization and codification are important commonplace features of practices. They provide evidence for the existence of a practice and also valuable data for studying understanding, detecting innovation and identifying authority. Encyclopaedias, self-help, teach yourself, recipe and guide books, etc. are artefacts which play an important part in the way practices operate and are coordinated, and they should be understood in the context of extensive and diversified processes of intermediation. As a consequence, practices are more or less tightly organized and coordinated. Strong coordination occurs in circumstances where instrumental production is involved, as with handicrafts or manufacture, and perhaps where there are clubs or professional associations of practitioners who recognize and covet the valuable resources (symbolic and material) at stake, especially if there are enforceable penalties for poor performance. *Per contra*, weak coordination is likely to occur in non-production and non-instrumental activities such as those associated with consumption, especially where activities are of low symbolic value or where component practices are themselves strongly but independently coordinated.

Notes

1 Postill (2010) identifies three phases in the development of the modern theory of practice, the first associated with Bourdieu and Giddens, a second with Schatzki, and the third with a range of attempts at the empirical application of the theory.
2 At the same time, there seems to have been a revival of interest in various forms of pragmatism, including pragmatist philosophy as a foundation for social science and sociological explanation (e.g. Kilpinen 2000; Baert 2005; Gross 2009; Martin 2011).
3 Few *definitions* of practices are directly helpful, but Schatzki (1996) offers a number of criteria by means of which we might identify such:

- Performances can be read as correct, acceptable or innovative (pp. 101–2).
- Performance 'expresses components of that practice's organization' (p. 104).
- People have, and share, words denoting the activity; practices (or performances?) are social by virtue of exhibiting 'coexistence with indefinitely many other people' (p. 105).
- A practice is more than the sum of individual doings and sayings involved.
- Some entity exists which is not simply in the minds of individuals: 'the organization of an integrative practice is out there in performances themselves not in the minds of actors' (p. 105).
- Performances are mutually intelligible among people who are exposed to the same activity as part of the same culture.

4 For example, several presentations at the 2011 European Sociological Association Conference, Research Network 5 'Sociology of consumption', Geneva, devoted considerable attention to this difficulty.
5 The dispersed practice of X-ing is a set of doings and sayings linked primarily, usually exclusively, by the understanding of X-ing. This understanding, in turn, normally has three components: (1) the ability to carry out acts of X-ing (e.g. describing, ordering, questioning), (2) the ability to identify and attribute X-ings, in both one's own and other's cases, and (3) the ability to prompt or respond to X-ings

(Schatzki 1996: 91).

6 'Teach Yourself' books being sold by Amazon at 6 March 2009 include: Weight Control through Diet and Exercise, Public Speaking, Book Keeping, Planets (part of a series Teach Yourself Science), Gaelic, Arabic, New Era Shorthand, Poker, Microsoft Office 2007, Anthropology, Humanism, Handwriting; Language instruction has been the topic of the most popular and numerous of the books.
7 They tend to be linear and not to present alternative ways of doing or judging; advice is thus typically not open to contestation, though this may be truer of older manuals than contemporary ones.
8 Eating, of course, draws also on many dispersed practices like remembering, reporting, seeing, tasting, imagining, etc.

References

Baert, P. (2005) *Philosophy of the Social Sciences: Towards Pragmatism*, Cambridge: Polity.
Bourdieu, P. (1978 [1972]) *Outline of a Theory of Practice*, Cambridge: Polity.
Bourdieu, P. (1984 [1979]) *Distinction: A Critique of the Social Judgment of Taste*, London: Routledge.
Bourdieu, P. and Wacquant, L. (1992) *An Invitation to Reflexive Sociology*, Cambridge: Polity Press.
Collins, H. (2010) *Tacit and Explicit Knowledge*, Chicago Il: Chicago University Press.
DeLanda, M. (2006) *A New Philosophy of Society: Assemblage Theory and Social Complexity*, London: Continuum.
Elias, N. (1969 [1939]) *The Civilizing Process, Vol. I, The History of Manners*, Oxford: Blackwell.
Ferguson P. (2004) *Accounting for Taste: The Triumph of French Cuisine*, Chicago: Chicago University Press.
Fine, G. (2010) 'The sociology of the local: action and its publics', *Sociological Theory*, 28(4): 355–76.
Giddens, A. (1984) *The Constitution of Society*, Cambridge: Polity.
Gross, N. (2009) 'A pragmatist theory of social mechanisms', *American Sociological Review*, 74: 358–79.
Kilpinen, E. (2000) 'The Enormous Fly-wheel of Society: Pragmatism's habitual conception of action and social theory', University of Helsinki, Sociology Department, Research Report 235.
Lizardo, O. (2009) 'Is a "special psychology" of practice possible?: from values and attitudes to embodied dispositions', *Theory and Psychology*, 19(6): 713–27.
McFall, E., du Gay, P. and Carter, S. (2008) *Conduct: Sociology and Social Worlds*, Manchester: Manchester University Press.
Martin, J. L. (2011) *The Explanation of Social Action*, Oxford: Oxford University Press.
McIntyre, A. (1985) *After Virtue*, second edition. London: Duckworth.
Ortner, S. (1984) 'Theory in anthropology since the Sixties', *Comparative Studies in Society and History*, 26: 126–66.
Petrini, C. (2001) *Slow Food: The Case for Taste*, New York: Columbia University Press.
Pinkard, S. (2009) *A Revolution in Taste: The Rise of French Cuisine*, Cambridge: Cambridge University Press.
Postill, J. (2010) 'Introduction: theorising media and practices', in B. Brauchler and J. Postill (eds), *Theorising Media and Practice*, New York: Berghahn Books.
Reckwitz, A. (2002) 'Toward a theory of social practices: a development in culturalist theorizing', *European Journal of Social Theory*, 5(2): 243–63.
Schatzki, T. (1996) *Social Practices: A Wittgensteinian Approach to Human Activity and the Social*, Cambridge: Cambridge University Press.
Schatzki, T. (2002) *The Site of the Social: A Philosophical Account of the Constitution of Social Life and Change*, Pennsylvania: Penn State Press.
Schatzki, T (2011) 'Practice theory' in D.Southerton (ed.) *Encyclopaedia of Cultural Consumption*, London: Sage.
Schatzki, T., Knorr Cetina, K. and von Savigny, E. (eds) (2001) *The Practice Turn in Contemporary Theory*, London: Routledge.
Trubek, A. (2000), *Haute Cuisine: How the French Invented the Culinary Profession*, Philadelphia, PA: University of Pennsylvania Press.

Tuan, Yi-Fu (2005) 'Pleasure of the proximate senses: eating, taste, and culture', in C. Korsmeyer (ed.) *The Taste Culture Reader: Experiencing Food and Drink*, Oxford: Berg.

Warde, A. (2004) 'La normalita del mangiare fuori', ('The normality of eating out'), *Rassegna Italiana di Sociologia*, (special issue on 'Sociology of Food') R Sassatelli (ed.), 45(4): 493–518.

Warde, A. (2005) 'Consumption and theories of practice', *Journal of Consumer Culture*, 5(2): 131–53.

Warde, A. (2009) 'Globalisation and the challenge of variety: a comparison of eating in Britain and France', in D. Gimlin and D. Inglis (eds) *The Globalisation of Food*, Oxford: Berg.

3 The edge of change

On the emergence, persistence, and dissolution of practices

Theodore Schatzki

This essay seeks to identify the locus and basic dimensions of change in social practices. I believe that changes in social practices are fundamental to changes in social life more broadly. This is because social phenomena are bundles of practices and material arrangements.

The account presented here embraces (1) general propositions about human activities, practices, and practice-arrangement bundles and (2) an enumeration of key aspects and components of the emergence, persistence, and dissolution of such bundles. This enumeration does not aim for completeness. Analyzing all significant aspects would be an immense task, besides which lists like this are open-ended, being the joint products of empirical investigation and thought informed by such investigation. As a result, they are open-ended. Note that the general propositions found in part one of my account, specify the locus and basic dimensions of the subject matter that empirical investigation and thought can further study.

The current essay declines to construct a general explanatory template, in this case, of the emergence, persistence, and dissolution of practice-arrangement bundles. Science's interest in explanation often combines with its drive toward simplicity to produce explanatory schemas that trace diverse changes to one or two phenomena. Part one of my account does something of this sort ontologically in uniting a broad range of phenomena under a single framework. The enumeration in part two, however, pulls in the opposite direction. In laying out different aspects and components of persistence and change, it offers a range of considerations that might, or might not, be pertinent in particular cases.

My discussion, I should add, heavily leans toward the theoretical. I usually like to substantialize theoretical ideas through the extensive development of empirical examples, but space considerations preclude that approach here. I hope, nonetheless, that sufficient examples are mentioned to make the theoretical package intelligible.

Endless happening

Part one lays out the ontology of activities, practices, and bundles that underlies my discussion (in part two) of aspects and components of emergence, persistence, and dissolution. I have developed this ontology at great length elsewhere (Schatzki 2010) and will only outline it here.

As I analyze things, activity or performance (activity is the performance of an action), is an event. To say that it is an event is to say that it happens. Many theorists will affirm that activity is an event, although they will disagree among themselves about what it is to be an event, what sort of event activity is, and what follows from activities being events. Most theorists will also insist on contrasting activity to 'mere occurrences' such as solar eclipses, chemical reactions, and waking from a state of hibernation, on the grounds that activities – and not other events – are intentional and/or voluntary events, which are carried out (by people). So construed, however, activities, and not just mere occurrences, are events.

Not all thinkers treat activity as an event. A prominent group of thinkers claims that human life (including human activity) is, most importantly, a process; a continuous flow or unfolding. An impressive cadre of thinkers have held this view, including William James (1952), Wilhelm Dilthey (1927, vol. 7), Henri Bergson (e.g., 1988), Edmund Husserl (1991), Alfred North Whitehead (1979), George Herbert Mead (1932) and, more recently, Gilles Deleuze (1988), Anthony Giddens (1979), and Tim Ingold (2000).[1] I agree that that human life is marked by continuity and a sense of continuousness. Continuity and continuousness, however, do not mark or result from the continuous unfolding of a stream of consciousness, life, or activity. They are, instead, reflections of the fact that a person is always doing something or other (so long, at least, as she is awake): a human life embraces a continuum of activity, or performance. This continuum, moreover, is broken into series of possibly overlapping activities: I cook the eggs, tidy the counter, and make a grocery list, while talking on the telephone. Any of the activity events in such a series might itself unfold in the sense of continuously occurring over a stretch of time. Taken together, however, these activity events do not add up to one continuous unfolding. Instead, they form a gapless series of overlapping events: each distinct, each a new beginning, and each concluded, interrupted, or forsaken.

A human life is, prominently, a continuum of activity. As a result, any state of affairs encompassing more than one person embraces a multiplicity of such continua. This is a basic fact about human life that is fundamental to the character of sociality and society.

Activities happen. Happening, however, is not equivalent to change. To happen is to take place, to occur, to become part of the inventory of what is. The performance of an action does not necessitate any more change than that the stock of events in the world has increased by one. In particular, a performance need not implicate further changes in social facts, phenomena, or events. In other words, an activity can just as easily maintain the world as alter it. In fact, this is the usual case. This observation differentiates accounts that treat activity as event from those that treat it as process or becoming. According to the latter accounts, every event counts as change. To paraphrase Deleuze: to become is to become different (Deleuze 1988).

Activity is an event. It is, more expansively, a temporalspatial event. Activity, is not, however, "temporalspatial" in the familiar sense of occurring in time and space understood as arenas or containers of some sort. Nor is it temporalspatial in the sense that configurations of time and space are built up from activities as forms of relational time and space. In describing activity as temporalspatial, I mean that a non-objective time and space, better, a non-objective timespace is a constitutive feature of activity. Activity is a temporalspatial event in the sense that a non-objective timespace makes human activity what it is.

The time component of activity timespace is what Heidegger called "temporality" (Heidegger 1978). Temporality is the future-present-past dimensionality of human activity. According to Heidegger's analysis, the future dimension of activity is coming toward something in acting, while the past dimension is starting or departing from something. To come toward something in acting is to act *for the sake of* that thing. In more conventional language, it is to act for an end and purpose(s). To start or depart from something in acting is to act *because* of that thing; more conventionally, it is to react to or act in the light of that thing; to be motivated in what one is doing.[2] The present dimension of activity is acting itself. In acting, a person comes toward that for which he acts and departs from that because of which he acts as he does. This is the temporality of activity. More conventionally expressed, the temporality of human activity is teleologically acting motivatedly.

The space component of activity timespace is what Heidegger called "spatiality." Spatiality is the world around in its pertinence to and involvement in activity. More specifically, spatiality is the distribution of places and paths through the material entities amid which people proceed: it is the material world amid which a person acts, housing interconnected places and paths for action, where a place is a place to do such and such and a path is a way from one place to another. The office in which I presently sit, for instance, houses places to write, to hold conversations, to put my briefcase, and the like that are anchored at desks, chairs, and tables. Spatiality is how the material world forms a setting for human activity. Activity inevitably transpires in a material world that it appropriates as its setting.

The pertinence of materiality to activity extends beyond the fact that performances happen amid material entities, attuned to the places and paths these entities anchor. People also react to events and states of affairs (the past dimension of activity) that befall or hold material entities. If the lights go out in my office, I might stand up and go investigate what's going on. Materiality can also fill out the ends and purposes for which people act (the future dimension of activity); in order to get the busted light switch repaired, I might call the university's physical plant. Beyond this, people, in acting, have to negotiate the physicality of the material world about them. They have to maneuver around things and know how to produce them or transform them into other things by shaping, fragmenting, connecting (etc.). They also need to know what can be done with what, and what will happen to objects under various conditions. Because of this, the physical-chemical properties of material entities are relevant to human activity (and social life) even if people know little, or have false ideas, about these properties.

The pertinence of materiality extends beyond this. A person does not just act within an immediate material setting; for instance, my office. This setting is connected with other material arrangements; for instance, those composing adjoining rooms, other floors of the building, the ground on which the building stands, the neighbourhood, the town etc. These configurations can be relevant to activity in my office. A cry emanating from an office across the hall might lead me to rush out of my office and across the hall. A chemical explosion at the electricity plant that powers the building might close down the plant, causing the lights in my office to be extinguished. Material arrangements form immense interconnected networks through which causal processes work, affecting both the arrangements themselves and the human activity that transpires amid them.

Materiality also bears on activity because the performance of most actions consists in the performance of bodily actions. Waving, for instance, consists in moving one's hand back and forth in certain circumstances. A bodily activity is a bodily doing or saying that a person can perform directly; that is, perform without doing something else. Almost all humans master a range of such doings and sayings; I call this range a person's "bodily repertoire." Most activities –including most mental activities – essentially consist of people carrying out bodily actions in certain circumstances. The circumstances include antecedent actions and events as well as results and consequents.

Activity is an event. It is a temporalspatial event. It is a temporalspatial event that (in addition to appropriating the material world as a setting where it happens) is exposed to wider material events and arrangements, linked to that setting, and it befalls someone who carries it out through his or her performance of bodily actions.

Almost all activity events are, in addition, occasions on which practices are carried out. Standing up to check out the light switch, like calling the physical plant and rushing across the hall when someone cries out, are moments of social practices. In performing them, I *ipso facto* carry out the practices involved. By a "social practice" I mean an organized, open-ended manifold of activities spread out over objective time and space. The performances that compose a practice are organized by items of four types: (1) practical understandings, (2) rules, (3) teleological structures, and (4) general understandings. By "practical understanding," I mean knowing which bodily actions to perform (in particular circumstances) in order to accomplish specific actions. "Rules" are formulated directives, admonishments, and edicts. A "teleological structure," meanwhile, is a range of prescribed or acceptable ends, coordinated with a range of prescribed or acceptable projects, together with actions to be carried out in order for those ends to be achieved. And "general understandings" are understandings or senses of general matters pertinent to goings-on in the practice. The performances that compose car repair practices, for instance, are organized by (1) practical understandings of what bodily actions to perform in carrying out such actions as rotating tires, timing engines, replacing tie rods, finding parts on-line, and the like, (2) rules about such matters as cleaning up, dealing with automobile dealer parts departments, and informing customers of impending problems, (3) prescribed and acceptable ends such as making money, satisfying customers, and staying in business, tied to projects and actions that are prescribed or can acceptably be carried out in pursuit of these ends, and (4) general understandings about, among other things, the beauty of a well-tuned automobile.

Most activities are elements of one or more practices. This means that most activities extend the manifolds of extant and past activities that compose one or more practice. Activities are also circumscribed by the organization of practices and, in turn, typically maintain these organizations. The idea that activity reinstates the structure that governs it is a familiar motif affirmed by diverse theorists, most notably, Anthony Giddens (1979), Pierre Bourdieu (1976), and Roy Bhaskar (1979). My account of how practice organization circumscribes activity differs considerably from theirs, but I will not discuss this topic presently.

Because it circumscribes activity, practice organization shapes the temporal and spatial dimensions of activity. That for the sake of which a person acts, that in reaction to or in whose light she does so, what she does, and the arrays of places and paths that both are anchored in the material things amid which she acts and help determine what she does, reflects the teleological structures, general understandings, and rules that organize practices. As they carry on their day, for example, the ends for which auto mechanics and auto shop managers act, like the events to which they react and the places and paths anchored in the arrangements that compose shop floors, offices, and kitchen areas, are sensitive to the organization of car repair practices. Generally speaking, the organization of a practice specifies common temporalities and spatialities for its participants, namely, those enjoined of participants. The organization of a practice also underlies participants' *shared* temporalities and spatialities. These common and shared temporalities and spatialities are joined by *orchestrations* of temporality and spatiality that are effected in the enactment of the practice, to form an interwoven timespace that runs through the practice (temporality or spatiality are orchestrated when the temporalities or spatialities of different persons' activities diverge but are not independent of one another).

I just adduced one way in which practices and material arrangements are linked: through arrangement-anchored, place-path arrays that are common to, and shared by, practice participants due to the organization of the practice. Practices and material arrangements are linked in other ways: (1) by virtue of activities that compose practices altering arrangements, (2) by way of activities reacting to events that befall these and other arrangements, (3) through causal relations among elements of arrangements affecting the progress of practices, (4) through material arrangements prefiguring the progress of practices,[3] (5) by virtue of the impossibility of carrying on certain practices in the absence of certain arrangements (e.g., car repair in the absence of cars and electronic instruments), (6) through the dissemination of certain arrangements as particular practices spread, and (7) through participants in particular practices making sense of the elements of arrangements in specific ways. Incidentally, (5) combined with (6) suggests that practices and arrangements constitute one another. Just as many practices would not exist or would look different if certain arrangements did not exist, so, too, would many arrangements not exist or take different forms were it not for certain practices.

Through links such as these, practices and arrangements form bundles. Elsewhere, I have suggested that social phenomena can be understood as slices or aspects of practice-arrangement bundles (or of confederations of these) (Schatzki, 2003). Activity events maintain or change such bundles in perpetuating or altering social practices, in appropriating or altering the arrangements linked to these practices, and in maintaining or changing the relations of practices to arrangements and to one another. Incidentally, examining the avenues through which practices entwine and arrangements connect, would lead the present discussion astray. In part because of this, my discussion below will largely abstract from these entwinements and connections.

Human activity is an event. It is a temporalspatial event. It is a temporalspatial event that both appropriates and intervenes in a material setting as a place where it takes place; that is exposed to and affected by the wider material world linked to that setting; that befalls someone who carries it out through the performance of bodily actions; that typically carries out and therewith perpetuates or alters one or more social practices; and that, thereby, perpetuates or alters not just the practice-arrangement bundles of which these practices are part, but also social phenomena (including the interwoven timespaces and practice organizations involved in the bundles and social phenomena). In turn, practices, arrangements, bundles, and social phenomena are the determining contexts in which activity happens.

I mentioned above that a human life encompasses a continuum of activity. Social states of affairs generally embrace multiple individuals and, thus, multiple performance continua. It is chiefly in the unfolding of multiple, interconnected performance continua – though also in the occurrence of events that befall material arrangements, especially those amid which people act – that social life abides and changes. Practices and social phenomena arise, persist, and dissolve through the twists and turns of activity.

The idea that human action is central, constitutively and causally, to social affairs has always been prominent in social thought. A substantial portion of thinkers who affirm this idea have been ontological or methodological individualists, for whom the centrality of action to social affairs is a reflection of the centrality of individual human beings to this. I converge with such thinkers in construing activity as an event that befalls individual people. This construal does not, however, accord individuals ontological supremacy. Elsewhere, I have discussed how individuals (in particular) their activities and mental conditions, depend on and presuppose social practices. Above, moreover, I intimated that human activity is inseparable from material arrangements, practice-arrangement bundles, and social phenomena. The fundamental locus of social life is not centred on individuals and their minds and actions, but on constellations of activity continua, practices, and material arrangements. Activities are one, albeit key, component of these complexes.

Emergence, persistence, and dissolution

According to the ontological picture just outlined, social stasis and change (the core theoretical topic of this book) centrally transpire in or through the temporal-spatial activity continua that inherently form parts of practice-arrangement bundles. The ebb and flow of established and new practices arises from the complex intercalation of activities, material arrangements, and practices.

When most social investigators speak of "practices," they mean configurations of actions and material entities, rather than organized activities alone. Because of this, I focus in the following on the emergence, persistence, and dissolution of practice-arrangement bundles. Since social phenomena are slices or aspects of such bundles, the focus is, at once, the emergence, persistence, and dissolution of social phenomena. I will also simplify matters by examining bundles and abstracting from the facts that bundles form confederations and that confederations form ever wider confederations that are ultimately coextensive with the objective spatial-temporal spread of social life.

Emergence

The emergence of a practice-arrangement bundle is the establishment of one or more organized activity manifolds that conjointly transpire amid a particular arrangement or set of similar arrangements. This process has multiple, combinable components.

The first component is the coalescence of the practice(s) involved; the existence of organized activities. There is no definitive criterion of when this has occurred. Regularity is no criterion (though it is often a clue), since the organized activities that compose a practice need not be regular. The coalescence of a practice instead involves some combination of (1) the emergence of common rules (explicit formulations) in the light of which actors proceed, (2) the crystallization of sets of prescribed or acceptable ends, tasks, and actions (however disputed particular norms might be or however indefinite the limits of the set might be), (3) the development of common practical understandings (the mark of which is its being intelligible to others that people perform particular bodily actions when carrying out a certain action), and (4) the distillation of common general understandings. New practices do not require new arrangements and can coalesce amid existing ones. The practice, for example, of dinner guests congregating in kitchens contrasts with prior practices, which frowned upon guests in kitchens, even though the arrangements amid which these practices proceed(ed) are highly similar.

The emergence of a practice-arrangement bundle can also be tied to the production and introduction of particular material entities and arrangements. A new bundle can emerge when the built environment is significantly altered or people colonize a further expanse or dimension of nature. When people simply extend existing practices into these new milieus, what results is more an evolution of extant bundles than the emergence of novel ones (especially if the only changes to the practices are the inevitable adjustments to new settings). The movement of settlers onto new land and the occupation of newly built urban and suburban structures exemplify possibilities. More significant changes ensue from the introduction of material innovations or the intentional redesign and reconstruction of built environments with an eye toward accommodating or enabling particular practices. An example of the former is the introduction of televisions into living rooms, which (in the US) ushered in evolved or new family practices such as gathering to watch shows and eating TV dinners. Other examples are the introduction of recycling receptacles and new family recycling and trash collection routines, and the introduction of cell phones and new practices of communication and coordination. The changes in prison design famously studied by Foucault, provide striking examples of built environments constructed to enable particular practices (Foucault 1979). The emergence of practice-arrangement bundles embraces changed practices, changed arrangements, or altered links between practices and arrangements.

When extant practices occupy new arrangements, but not only then, small changes in practices can occur, the accumulation of which eventuates in evolved or new practices. The evolution of cooking practices as technologies and techniques develop is an example, though some features of cooking change glacially.[4] The migration of practices into new arrangements can also bring about transformed practices that thenceforth evolve differently from the versions of the practices that did not migrate. Such a transformation is an emergence. Colonial history offers endless examples, in which, say, a governmental practice is brought from the colonizing country to the colony, where, meshed with new practices and arrangements, its subsequent evolution diverges from the evolution of the mother practice.

A different sort of emergence occurs when a practice-arrangement bundle bifurcates and its descendants evolve separately from one another. New sports often emerge in this way: to smallish changes that commence a separate evolution, additional changes accrue. The emergence of American football took this form, arising from English rugby football and "mob" football.

Bundles can also hybridize,[5] yielding new (though not necessarily more complex) bundles. Mergers of firms or academic units provide different sorts of examples. Conquering armies (also) traditionally left such hybridizations in their wake, as the conquerers' bundles combined with those of the conquered. More peaceful examples of the same process are labs joining to pursue a single research program and increased connections among world cuisines leaving hybrid cooking-kitchen bundles in their wake.

A different sort of emergence involves what Deleuze and Guattari call a "line of flight" (Deleuze and Guattari 1987: *passim*; Deleuze 1997: 187). A line of flight is the path taken by a person, artefact, organism, or thing that flees a given bundle and thereby possibly contributes to that bundle's dissolution. New bundles sometimes coalesce around such entities. Many small businesses, for example, depend on particular individuals. When such an individual departs and starts a new firm, she follows a line of flight.

A practice-arrangement bundle rarely emerges instantaneously. One reason for this is that a bundle requires a reasonably stable organization in order to exist; absent this organization and what exists is continual metamorphosis, not a new bundle. Periods of ferment or transition often precede the emergence of bundles. Such protean periods sometimes occur, for example, when technological devices are introduced. Periods of ferment or transition give way to new bundles as people's activities, often without their knowledge, come to be governed by common rules, teleological structures, and general understandings. Generally speaking, because (1) practices are carried out by multiple people, (2) activities form sequences, and (3) people's reactions to events and circumstances differ, the emergence of commonalities takes time. Social life, as a result, often exhibits pockets of experimentation, uncertainty, and convergence amid stability and gradual evolution.

Another requirement for emergence is the development of new practical understandings. One mark of the existence of a practice or bundle is people knowing what bodily actions to perform in order to carry out the actions, performances of which are elements of the practices concerned. When changes in practices or bundles include new projects and actions, or new ways of carrying out existing projects, the practical understandings (and bodily repertoires) of participants must change in order for new or evolved practices and bundles to stabilize. Such a development occurs whenever, say, new procedures are introduced into production practices or technological advances lead to reorganizations of the work place (a dramatic example is Taylorism). New or evolving practices of greeting and communication encompass similar demands. It can also become necessary to learn new ways of carrying out familiar actions, for example, turning on the television as the numbers of electronic devices in family and living rooms increase. These necessities, and people's ability or inability to satisfy them, can have repercussions for group dynamics and for social stratification and differentiation.

Persistence

Identity is often conceived, in philosophy and everyday life, as the perpetuation of the same; that is, as the absence of change. This conception of identity does not apply to the persistence of practice-arrangement bundles over time; indeed, it does not apply to anything, but that is a different story. The persistence of a bundle over time is, instead, a kind of unity in difference. The subsistence of a bundle is compatible with the multiplication and metamorphosis of its activities, alterations of its arrangements (including connections among arrangements), transformations of interwoven timespaces and practice organizations, and changes in how practices link to one another and to arrangements. What is required for a bundle to persist through these kinds of changes is that the changes generally hang together and that they are neither too frequent nor too large. Persistent or massive changes of these sorts yield bundles whose composition so greatly differs from those of predecessors that they are no longer the same bundles; unconnected changes in these registers, meanwhile, suggest that a bundle has either ceased existing or that it never existed at all. Bundles persist over time when the changes they undergo are limited, stepwise, accumulating, and occur amid general continuity in activities, arrangements, interwoven timespaces, practice organizations, entwined practices, and links between practices and arrangements.

I define 'persistence' as the same bundle existing before and after change. There are two major kinds of persistence: stability and evolution. Stability exists when changes are minute, isolated, and non-ramifying. Evolution occurs when bundles survive larger or multiplying changes.[6] Dissolution contrasts with stability and evolution. It happens when changes are large, disruptive, or cascading.

A practice persists only if the activities that compose it keep happening. In the absence of such performances, the practice ceases to exist. This dependence of practices on activities implies particular relations between past and present. In particular, it entails that a practice that has persisted sometime in the past (including the immediate past) continues to persist only if and when activities that help compose it happen again. Before such activities happen, it is indeterminate whether the practice still exists or will continue. In theory, long stretches of objective time can pass – without prejudice to the persistence of a practice – before an activity happens through which it is perpetuated. As long as the pertinent material arrangements exist and the activity manifold, interwoven timespaces, and organization involved can be extended or realized, activities might happen that maintain the practice in existence. Information storage in the form of memory, writing, electronic depositories, and the like is crucial to the persistence of practices over temporal gaps among its activities. To the extent that human memory is required, practices cannot persist over time gaps longer than the lives of individuals or the three to four generational link of storytelling. Persistence across temporal gaps longer than this are infeasible at this point in human history. The persistence of practices also requires the endurance of material arrangements or their elements.

Memory points to a second key feature of stable practice-arrangement bundles, namely, the stabilization of the practical understandings through which people perform certain bodily actions when carrying out the actions that compose particular practices. This is a key element of the information storage (in this case, memory) required for bundles to persist. People's bodily repertoires, which are coordinated with these practical understandings, must also be stable. Of course, understandings and repertoires can metamorphose along with changes – smaller and larger – in the activities, tasks, and material entities embraced in persisting bundles. Still, understandings and repertoires tend toward inertia; indeed, their inertia is crucial to the pervasive persistence of bundles, a conservatism which is inherent to social life. This conservatism is all the more extensive given that the same practical understandings and bodily repertoires can underlie multiple bundles.[7] This inertia also contributes toward the fact that it takes time for bundles to emerge, as well as to the availability of activities and

The edge of change 41

entities to migrate among, or to be appropriated by, bundles. This availability is a flip side of the difficulties older workers face when they lose jobs that they have held for decades.

A further feature of the endurance of a bundle is the maintenance of its temporalspatial infrastructures. Temporalspatial infrastructures are central to a stable bundle because interwoven timespaces form the interconnectedness of the constitutive dimensions of the bundle's activities. Such infrastructures are circumscribed by practice organizations, anchored in and underlain by material arrangements, and enabled by practical understandings. A bundle's temporalspatial infrastructures are preserved by the very activities whose interconnected dimensions they are. What I mean is that common, shared, and orchestrated timespaces persist by virtue of participants' activities extending them: their maintenance is an effect of these activities.

The organizations of practices are likewise maintained by the activities that uphold them. Arrays of ends, tasks, and actions that people are enjoined to or can acceptably pursue, endure only if people's activities conform to them. Consquently, organizations are maintained by the activities that compose the practices they organize. Dialogue, admonition, persuasion, argument, threats, and the like can be causally responsible for the persistence of organizations or interwoven timespaces. But unless people actually conform to the organization or extend the infrastructure, it ceases. This figure of thought is familiar from the work of Giddens, Bourdieu, and Bhaskar, though as stated above – and as suggested by my remark about causal responsibility – my account of how activities come to conform to practice organization diverges from theirs.

A final component of stable bundles is the continuing existence of the same, or similar, material arrangements. The arrangements that help compose many a persisting bundle change only trivially or occasionally. The arrangements composing an academic department or a country store are often like this. Bundles can also persist through significant changes in arrangements. Firms and governments survive when new buildings are occupied and old ones are abandoned or destroyed. Of course, changes in arrangements can spark or accompany larger changes in bundles, including in their relations with one another and in relations among their constituent practices. This process can result in evolved or new bundles. The construction of a lake in a park leads to evolved recreational bundles there, just as the construction of recycling facilities in a town where none existed before leads to new activities amid old and new arrangements.

As indicated, bundles evolve when they evince accelerated, pervasive, or nonramifying changes in interwoven timespaces, practice organizations, practical understandings, material arrangements, relations among practices, causal relations within bundles, prefigurational relations, and constitutive relations between activities and entities. As indicated, however, the difference between stability and evolution is not definite. It is a difference between small, fewer, and nonramifying changes, on the one hand, and large, frequent, or multiplying changes on the other. There is no metric, moreover, with which, say, non-ramifying and multiplying changes on the one side and disruptive changes on the other, can be cleanly demarcated. The complexity of bundles typically proscribes this. Sometimes, of course, the difference between stability and evolution is straightforward. Baseball, for instance, is the same game it was ninety years ago, though it has evolved and not remained stable. The use of plates, glasses, and cutlery in Western countries is also the same as it was ninety years ago, though it is stable and has scarcely evolved. Have, however, US military ground combat tactics remained stable or evolved in recent years? The answer is not so apparent (not, of course, that an answer must be given; there may be no pressing need to pose the question as formulated).

An important kind of evolution embraces developments of more complex bundles through alterations of the relations among practices, the connections between arrangements, and the conjunction of these two processes. Corporate reorganizations, for example, embrace new links between the practices and new connections among the arrangements that compose the complex bundle that is the corporation. Practices and arrangements can also be added to, and thereby augment, existing bundles evolutionarily. Firms, academic units, and armies can all grow by absorbing additional practices and arrangements.

Something that bears much more on evolving than on stable bundles is the incursion of causality from outside the bundle. Generally speaking, changes can be generated within a bundle or induced or caused from outside it. Stability does not require that external change be largely absent since externally-caused or -induced changes in activities, arrangements, timespaces, organizations, and so on can be small or insignificant. External phenomena, however, can cause or induce large or significant changes, and do so quickly. Climatic, geological, and military events and processes provide dramatic examples. The fact that bundles are interconnected ensures that the world steadily serves up apprehended and unapprehended events and phenomena that can induce or cause significant changes in bundles: political events, technological advances, economic developments, microbial migrations, cultural trends, sports feats and extravaganzas, and unfamiliar religions, customs, and cuisines. The growing interconnectedness of humanity ensures the increasing frequency of externally wrought change (though it also constantly rewrites lines among bundles).

Dissolution

My comments on dissolution only scratch the surface of this complex phenomenon. Bundles dissolve when overwhelming, frequent, or large-scale changes occur to them. Dissolution does not equal destruction. External causes, such as the climatic, geological, military, and also biological ones just mentioned, often destroy bundles by eliminating the material locations needed to continue them, killing the humans required to perform them, or inducing massive responses from people that result in the abandonment of extant bundles or the emergence of new ones. More often, however, dissolution is a matter of smooth development from predecessors of bundles that embrace large, rapid, or cascading changes. Such cases are ones of linked simultaneous dissolution and emergence.

The edge of change 43

A dramatic example of this process is the ramp-up of the American rocket and space program that followed the Soviet launch of the Sputnik satellite: the changes were so great and accelerated that a new space program quickly emerged from the existing one. Contemporary automobile developments present another example. Hybrid cars look and are operated much like internal combustion cars. Although driving them is different, the bundle formed by driving practices, hybrid automobiles, and street arrangements is not new. Electric automobiles, by contrast, will require larger changes in practices and design and, when more affordable and widely available, will be part of a new bundle of driving practices, automobiles, and supporting arrangements.

Stability and evolution exhibit a range of combinations of internally and externally generated changes. By contrast, dissolution (and often emergence) almost always results from external inducement or incursion. External incursions include interruptions in material flows that are required for the sustenance of bundles. Agricultural bundles, for instance, often dissolve and are replaced by alternatives when the hydrological, biological, or climatic flows through the local ecosystem cease or are significantly altered.

An important way that bundles conjointly dissolve and emerge is through the selection of new ends for a bundle's practices, thus through changed interwoven timespaces and practice organizations. The ends that people should, or may, acceptably pursue anchor teleological orderings of practices; together with purposes, they also provide points of reference for the construction and set-up of the material world. Changes in high level ends – as, for example, when government or company policies radically change direction – are usually accompanied by new congeries of tasks and purposes (new teleological structures), new spatialities anchored in probably altered material arrangements, changed practical understandings, altered arrangements, and new links of practices to arrangements and to one another. It is in part because so much hangs together with them that high level ends do not change often in social life. When they do, however, old bundles dissolve and new ones begin. An interesting variation on this theme is the loss of belief in governing ends: when this happens, extant practices and the bundles they help compose become pointless in the eyes of practitioners or members. Because loss of belief in governing ends is rarely a rapid or painless process, the dissolution attendant to its occurrence usually requires time.

Dissolution can also result from hybridization and bifurcation. When two bundles join to form a third, the original bundles dissolve and are replaced by a new one. Similarly, when one bundle absorbs another, the latter usually dissolves, or at least, evolves dramatically. Business mergers provide fine examples of the range of possibilities. Bifurcations, too, can yield dissolved bundles, though they can also involve one bundle branching off from another and both continuing apace. The development of hybrid automobiles can be viewed as an example.

A final, curious, cause or dimension of dissolution is loss or irrelevance of practical understanding in changing circumstances. A possible example is handwriting among today's youth. It is not too far-fetched to say that along with (and as an accompanying cause) of changing communication practices, youth today is slowly losing the practical understanding involved in handwritten communication.

Conclusion

I have suggested that the emergence, stability, evolution, and dissolution of practice-arrangement bundles arises from temporalspatial activity events, in conjunction with events that befall either the material arrangements amid which activities happen, or further arrangements connected with these. My discussion has charted features and dimensions of change in bundles so conceived. This discussion, taken by itself, cannot offer specific prescriptions about how to effect particular changes, including sustainable responses to climate change. It does, however, identify loci and forms that change can take. It is reasonable, moreover, to call on individuals and organizations to work toward implementing or effecting such changes in the domains over which they exert some measure of governance or influence. The following loci and forms of change are pertinent to responding to climate change and are measures that individuals and organizations could consider:

1 Working to change the ends that are acceptable or prescribed in practices, the overall ends that govern what participants do.
2 Funding research into and producing energy-saving materials and devices for integration into extant practice-arrangement bundles.
3 Piloting the use of these alternative material arrangements and exploring how existing practices react to, appropriate, and hybridize with them.
4 Recruiting people to energy-efficient or sustainable bundles by (for example) educating people about dangers, encouraging new understandings of dwellings and workplaces as arrays of places to perform ecologically conscious actions, making it easier to experiment with or adopt promising bundles, and encouraging damaging ones to be abandoned.
5 Facilitating the development of sustainable bundles, for example, by financing "lines of flight" for individuals and entities that can form condensation seeds for energy-saving or sustainable bundles.
6 Researching and publicizing exemplars of lives and bundles that are already sustainable.
7 Adopting and exemplifying alternative lifestyles and technological bundles.

Notes

1 For uses of this idea in analyses of social and historical affairs, see Dilthey (1927), Mead (1932), Giddens (1979), Massey (2005), Abbot (2001), and Halbwachs (1980).
2 Compare Schutz's conception of in-order-to and because motives in 'Choosing Among Projects of Action', in Schutz (1962).
3 By "prefiguring," I mean qualifying paths of action, not as possible or not possible, but on an indefinite range of such registers as easy and hard, obvious and obscure, tiring and enlivening, short and long, etc.

4 For an outstanding treatment of elements of economy and society that change very slowly, see Braudel (1973).
5 For discussion of bifurcation and hybridization in experimental systems in science (a type of practice-arrangement bundle), see Rheinberger (1997).
6 Over a sufficient period of time, all persistence is evolution and all evolution yields dissolution.
7 This fact points to the importance of training in human life in the sense of Wittgenstein's *Abrichtung* (1958: para. 5) or Lefebvre's *dressage*: the disciplinary regimentation of how the body goes on. See Lefebvre (2004: 39). A similar idea is also, of course, found in Foucault, e.g., Foucault (1979).

Bibliography

Abbott, A. (2001) 'Temporality and Process in Social Life' in A. Abbott (ed.) *Time Matters: On Theory and Method*, Chicago: University of Chicago Press.

Bergson, H. (1988 [1896]) *Matter and Memory*, trans. Nancy Margaret Paul and W. Scott Palmer, New York: Zone Books.

Bhaskar, R. (1979) *The Possibility of Naturalism*, Atlantic Highlands: Humanities Press.

Bourdieu, P. (1976 [1972]) *Outline of a Theory of Practice*, trans. Richard Nice, Cambridge: Cambridge University Press.

Braudel, F. (1973) *Capitalism and Material Life, 1400–1800*, trans. Miriam Kochan, London: Weidenfeld and Nicolson.

Deleuze, G. (1988 [1966]) *Bergsonism*, trans. Hugh Tomlinson and Barbara Habberjam, New York: Zone.

Deleuze, G. (1997 [1994]) 'Desire and Pleasure' in A. Davidson (ed.) *Foucault and His Interlocutors*, trans. Daniel W. Mith, Chicago: University of Chicago Press, pp. 183–92.

Deleuze, G. and Guattari, F. (1987 [1980]) *A Thousand Plateaus: Capitalism and Schizophrenia*, trans Brian Massumi, Minneapolis: University of Minnesota Press.

Dilthey, W. (1927) *Gesammelte Schriften*, vols. 1 and 7, Leipzig: B.B. Teubner.

Foucault, M. (1979) *Discipline and Punish*, trans. Alan Sheridan, New York: Vintage.

Giddens, A. (1979) *Central Problems in Social Theory*, Berkeley: University of California Press.

Halbwachs, M. (1980 [1950]) *The Collective Memory*, trans. Francis J. Potter Jr., New York: Harper and Row.

Heidegger, M. (1978 [1927]) *Being and Time*, trans. John Macquarrie and Edward Robinson: Oxford, Blackwell.

Husserl, E. (1991 [1893]) *On the Phenomenology of the Consciousness of Internal Time*, trans. J.B. Brough, Dordrecht: Kluwer.

Ingold, T. (2000) *The Perception of the Environment: Essays in Livelihood, Dwelling and Skill*, London: Routledge.

James, W. (1952 [1890]) *Principles of Psychology*, Chicago: Encyclopedia Britannica.

Lefebvre, H. (2004) *Rhythmanalysis: Space, Time and Everyday Life*, trans. Stuart Elden and Gerald Moore, London: Continuum.

Massey, D. (2005) *For Space*, London: Sage.

Mead, G.H. (1932) *The Philosophy of the Present*, Chicago: University of Chicago Press.

Rheinberger, H. (1997) *Toward a History of Epistemic Things: Synthesizing Proteins in the Test Tube*, Stanford: Stanford University Press.

Schatzki, T. (2003) 'A New Societist Social Ontology', *Philosophy of the Social Sciences*, 33(2): 174–202.

Schatzki, T. (2010) *The Timespace of Human Activity: Performance, Society, and History as Indeterminate Teleological Events*, Lanham, Md: Lexington Books.

Schutz, A. (1962) 'The Problem of Social Reality' in M. Natanson (ed.) *Collected Papers I*, The Hague: Martinus Nijhoff.

Whitehead, A.N. (1979 [1929]) *Process and Reality*, second edition, Chicago: The Free Press.

Wittgenstein, L. (1958 [1953]) *Philosophical Investigations*, third edition, trans. G.E.M. Anscombe, New York: Macmillan.

Part II
The materials of practice

4 Transitions in the wrong direction?
Digital technologies and daily life

Inge Røpke and Toke H. Christensen

Introduction

The environmental implications of information and communication technology (ICT) have been the subject of study since the early 1990s, primarily with a focus on energy impacts. Other environmental impacts – for instance, related to extraction of raw materials, chemical use and waste handling – are highlighted in various studies (Kuehr and Williams 2003; Hilty 2008), but it is the energy impact that is most often discussed. In this chapter, we complement existing research by developing a perspective in which everyday life takes centre stage and in which we focus on the energy implications of changes in the temporal and spatial organisation of everyday practices. Our ambition is not to suggest new ways of assessing the net impact of ICT, but to develop ideas that will be useful in avoiding the negative features of ICT development and encouraging the positive aspects.

One strand of existing research deals with the direct impact of ICT equipment on electricity consumption, often in relation to standby electricity use. From the late 1980s, offices were increasingly seen as energy-consuming workplaces. At around this time, the rapid increase in standby consumption in households also appeared on the agenda (Sandberg 1993). Since then, residential electricity consumption related to ICT (including consumer electronics) has escalated further. In their report on *Gadgets and Gigawatts* (IEA 2009), the International Energy Agency claim that the global residential electricity consumed by ICT equipment grew by nearly seven per cent, per annum, between 1990 and 2008. Even allowing for anticipated improvements in energy efficiency, consumption from electronics is set to increase by 250 per cent by 2030 (IEA 2009: 237). In our own work, we have suggested that ICT can be viewed as a new round of household electrification (Røpke *et al.* 2010).

In addition to its direct impact on electricity consumption, the use of ICT at home, and in other settings, gives rise to energy consumption in relation to the production of equipment and the running of the infrastructure, such as server parks and sending masts. These effects are much less researched, but some studies indicate considerable impacts (Hilty 2008; The Climate Group 2008; Willum 2008).

Other researchers focus on the environmental consequences of ICT use in various economic domains, including the positive potential for environmental improvements. Erdmann and Hilty identify two 'green ICT waves' of empirical studies. The first was motivated by the environmental implications of the rise of the internet and the 'new economy' from the late 1990s to the early 2000s; and the second followed the IPCC's fourth report in 2007, which encouraged studies of the potential for ICT to reduce GHG emissions (Erdmann and Hilty 2010). The field includes both micro- and macro-level case studies covering domains like process automation, smart grids (and other applications in the energy sector), energy management in buildings and intelligent transport systems. Since these studies are extremely complex and carried out in many different ways, it is not surprising that their conclusions are diverse (Yi and Thomas 2007; Erdmann and Hilty 2010).

The macro models used for assessing the environmental impacts of ICT often distinguish between first-, second- and third-order effects (Hilty 2008; OECD 2010):

- First-order effects: environmental impacts related directly to the life cycle of ICT hardware, including the production, use, recycling and disposal of ICT.
- Second-order effects: environmental impacts due to the applications of ICT that have the power to change the processes of production, transport and consumption.
- Third-order effects: environmental impacts related to the medium- or long-term adaptations of behaviour and economic structures that follow from the availability of ICT and the services it provides. Rebound effects emerging from efficiency gains can be included in this category.

In general, the sum of the first-order effects is negative, while the net impact of the second-order and third-order effects, respectively, may be either positive or negative. The results from various studies differ, for instance, depending on the extent to which rebound effects are taken into account. Several recent studies in the 'second green wave' provide optimistic results, because they tend to concentrate on ICT applications that produce environmental gains and focus on second-order effects, ignoring first-order and/or third-order effects (The Climate Group 2008; Erdmann and Hilty 2010: 833).

Although the two strands of studies together cover the energy impacts of ICT quite extensively, we find that the treatment of the energy impacts of integrating ICT in everyday life is still inadequate. Consumers figure in studies of a few selected domains such as teleshopping, telecommuting, mobile work and a few cases of virtual goods (dematerialisation), but a broader account of ICT in everyday life is lacking.

This chapter integrates and reviews findings from our qualitative studies of the use of ICT in Danish households: first, a study conducted in 2004–2006 on families' use of ICT in an everyday-life context, previously reported in

Christensen (2008, 2009); and second, a study conducted in 2007–2008 on the implications of households' ICT use for energy consumption (Jensen et al. 2009; Røpke et al. 2010). These insights are introduced with reference to a theoretical framework that is designed to pinpoint features that make everyday life, and the use of ICT within it, more or less resource-demanding, using energy impacts as the main indicator. Having outlined this framework, we concentrate on how ICTs co-develop with these energy-demanding features of everyday life. Finally, we discuss the insights that such an approach provides and its relevance for initiatives to promote environmentally positive aspects of ICT.

Linking everyday life and energy demand

Our theoretical framework is inspired by the 'practice turn' that has swept through the social sciences in recent years (Schatzki et al. 2001). The philosophers Schatzki and Reckwitz have contributed to the formulation of a coherent approach to the analysis of practice (Schatzki 1996, 2002; Reckwitz 2002), which has been brought into analyses of consumption and everyday life by several authors including Warde (2005), and Shove and Pantzar (2005) (see review in Røpke 2009). Issues of time, space and practice have recently come into focus reflected, for instance, in a new anthology on time, consumption and everyday life (Shove et al. 2009) along with renewed interest in earlier contributions from a time geography perspective, as developed by Hägerstrand (1985), Pred (1981) and others. Since temporal and spatial aspects are particularly important in relation to the use of ICT in everyday life, our theoretical framework draws on a combination of practice theory and time geography. We present the framework as a series of theses:

(1) Practice theory is based on the idea that in the continual flow of activities it is possible to identify clusters of activities where coordination and interdependence make it meaningful for practitioners to conceive of them as entities (Schatzki 2002). An organised set of activities is seen as a coordinated entity when it is recognisable across time and space: a *practice* is both a relatively enduring and relatively recognisable entity (Shove et al. 2007: 71). Practices only exist when they are enacted, and this enactment reproduces and transforms the recognisable entity over time. Practices are considered to be the basic ontological units for analysis; they constitute individual actions and create social structures and institutions.

When a practice is performed, practitioners make linkages between a diverse set of heterogeneous elements that configure the practice. For use in empirical investigations, Shove and Pantzar (2005) summarised these elements as material, meaning and competence, or in other terms: equipment, images and skills (see illustration in Figure 4.1). For the purpose of this paper, the main point is that the connection between everyday life and the environment occurs through practices. In their everyday life, people cook, eat, sleep, take care of their children, play football and work (which covers a variety of practices). Consumption is an

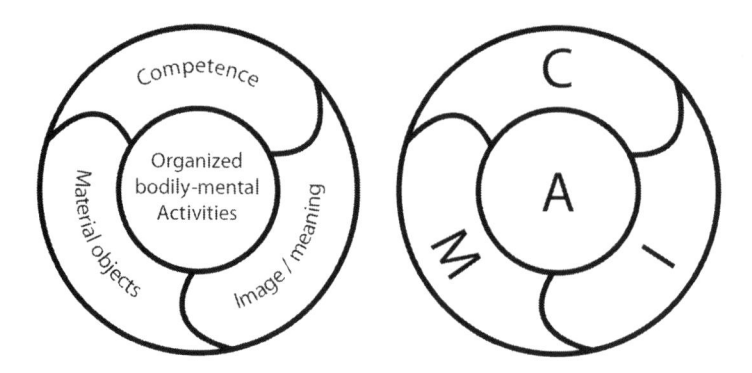

Figure 4.1 A practice as a configuration of competence, material and image.

aspect of practices, because the performance of a practice usually requires the use of artefacts, such as tools, materials and infrastructures. From an environmental perspective, the point is that the *use of resources* always takes place in relation to social practices.

(2) Since resources are mobilised through practices, the practices people combine in their everyday lives determine their environmental impacts. In a given society, some overall trends in the combination of practices depend on the *social and material framework* that has been established through previous practices. Figure 4.2 illustrates how people's 'performances' reproduce and transform webs of social and material structures that frame present and future practices. These structures include schooling and education, jobs in the formal economy, establishing a family, living in buildings, buying goods in shops, using

Figure 4.2 The interplay between practices and the social and material framework.

means of transport and so on. The social and material framework implies many restrictions and guidelines for the combination and enactment of practices in time and space.

(3) In addition, the individual, throughout his or her life, establishes a more *specific framework* within which practices are combined. Through a combination of possibilities (related to the individual's social background and inherited characteristics), choices and coincidences, a path-dependent biography is created: when a person has children, acquires a dog or buys a house, a private framework is created. To a great degree, the private framework is also influenced by the norms in a given society, and by the social group a person identifies with and has the economic possibility to belong to, but again with room for individuality. Sometimes people actively decide to shape their private framework in ways that restrict their choices and imply strong everyday priorities, such as having a dog that requires regular attention (Wilk 2009).

(4) The combination of collective and private frameworks defines a number of 'projects' in everyday life (Pred 1981). For instance, establishing a family forms an important part of a person's framework, which then defines the project of maintaining family relations. Similarly, having a job, a group of friends, a dog, a house or a garden, define other projects that many people consider to be central, and the projects may be interwoven – renovating the house can also be about building the family. A project is a complex of practices necessary to complete an intention, and it can be defined either by individuals or within an institutional context. It may be seen as a sort of meta-practice to which several 'sub-practices' relate. Usually, a number of practices are carried out in order to fulfil the aim of the project, but the practices are not necessarily bound to one project. Often a practice contributes to several projects – for instance, eating a meal may contribute to both the project of keeping fit and the project of maintaining family relations. These different projects are reflected in the meanings that are attached to and constitute the practice.

Practices may also be related to each other without necessarily being bound together through projects. Pantzar and Shove (2010) thus distinguish between 'bundles' and 'complexes' of practices, where a bundle is characterised by the co-existence of two or more practices that are minimally related (for example, by being co-located), and a complex denotes a situation of co-dependence in which practices are closely related and mutually dependent: 'The difference between bundles and complexes of practice has to do with the intensity and character of the links involved' (Pantzar and Shove 2010: 26).

(5) Everyday life unfolds in *time and space*: each individual follows a path in time and space, carrying out practices that take time and take place in space (Pred 1981; Hägerstrand 1985). It is a challenge to manage effective participation in practices within the limitations set by time and space, by institutionally defined projects and by the need for coupling one's own path with the paths of others, as is required to perform certain practices. Time resources are finite; it is impossible to be in more than one place at a time (the principle of indivisibility), and it takes time to move in space. When practices involve other people and

material objects, they depend on the coupling and uncoupling of the paths of all the 'partners', implying so-called coupling constraints. Projects function as organising devices that recruit participants and coordinate their paths. Coordination is also eased by 'pockets of local order' – that is, places like homes, workplaces and schools, where the elements needed for the performance of specific practices are located in accessible ways (Ellegård and Vilhelmson 2004). Figure 4.3 illustrates the movements in time and space.

The horizontal plane illustrates space, and time is measured vertically, from morning at the bottom to evening at the top. Each person (a–d) moves in time and space along a 'pipe', lined with time intervals. Parents A and B go to work, child C to school. Later, parent B meets child C at the sports hall, and parent A meets friend D in the shopping centre. Everybody returns home.

Collective time structures – like fixed working times, opening hours for shops, fixed meal times and recurrent television programmes – make coordination easier. Modern societies still have collective time structures, but much more flexibility has been introduced, which increases the task of coordination (Southerton 2009).

The temporal qualities of the totality of daily practices that make up individuals' and families' everyday life can be described metaphorically as a timescape, a concept originally coined by Adam: 'Where other scapes such as landscapes, cityscapes and seascapes mark the spatial features of the past and present activities and interactions of organisms and matter, timescapes emphasise their rhythmicities, their timings and tempos, their changes and contingencies' (Adam 1998: 10).

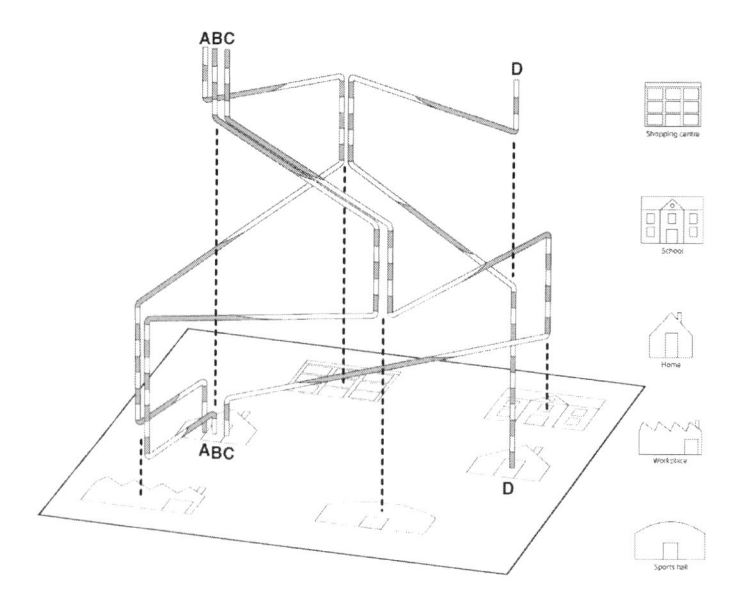

Figure 4.3 Movements in time and space.

(6) Everyday life is permeated with *relations of power and dominance*, reflected in unequal access to benefits as well as exposure to burdens and risks. Likewise, people have different degrees of influence over the projects that take up both their own time and the time of others. Pred (1981) emphasises the importance of power relations by raising the question: who has the power to coordinate the paths of others in relation to which projects? Those who hold power and authority within institutions are able to define projects, and the projects of dominant institutions in society tend to take precedence in terms of time-allocation and scheduling. For instance, most people give high priority to the demands from workplaces and schools. Also, at the household level, issues of dominance – related for instance to gender and age – are aspects of the complex family dynamics that are relevant for the relative priority of projects. As features of the social and material framework referred to in thesis 2, relations of power and dominance are continuously changed through the unfolding of practices.

(7) In spite of path-dependent framework conditions, collective time structures, 'pockets of local order' and relations of power and dominance, people still face a considerable *complexity* in everyday life. Most people have many projects and a high level of ambition with regard to their implementation, and to manage their participation in projects and practices and to make priorities, people tend to rely on *routines* – these being practices that are recurrently reproduced (Wilk 2009).

The link between everyday life and the use of resources can be spelled out in more detail with reference to these theses. As stated in the first thesis, the link is established through practices: the use of resources consequently figures as an aspect of practices. Focusing on energy as an example, the prevailing *energy-intensity of everyday life* can be said to depend on the following factors:

- the energy-intensity of each specific practice;
- the combination of practices taken up by practitioners;
- the number of practices practitioners are able to carry out within the temporal constraints;
- the extent of the space covered by practitioners when carrying out the practices.

Over time, each of these factors changes in response to the dynamic character of the social and material framework of everyday life; thus, new projects emerge and others disappear, for example, due to socio-technical innovation in business, government and civil society.

ICTs have consequently co-developed alongside changing practices and projects in everyday life. In the next part of the chapter, we consider the impacts of ICT on the energy-intensity of practices and on the temporal and spatial coordination of practices and projects. But first, a few words on how ICTs are integrated into practices.

Integrating ICT into practices

ICTs have a long history within industries such as telecommunications, recorded music, film, radio, television and office equipment; but the concept of ICT is of a more recent origin, related to the merger of technologies for communication, broadcasting and data processing. The basis for the merger was the emergence of the transistor and later the microchip, which made it possible to install an ever-increasing number of transistors in a very limited space. This miniaturisation enabled the inclusion of advanced data-processing facilities for monitoring, management and manipulation in a multitude of products, as well as the development of the general-purpose personal computer and the infrastructure of the internet. ICTs are now defined as generic or general-purpose technologies that can be used for all kinds of activities that involve the acquisition, storage, processing and distribution of information (Steinmueller 2007).

The general applicability of ICT implies that many consumer products, like washing machines and cars, are now equipped with various kinds of programming functions; but as a category of consumer goods, the concept of ICT is usually reserved for products and services related to entertainment (consumer electronics like television, radio, music and games), communication and administrative tasks like word-processing and calculation. In the following, we focus mainly on the use of computers, mobile phones and the internet in everyday life. The interpretive flexibility of these ICTs is very wide, which makes it interesting to discover how they are integrated into a variety of everyday practices.

In our study of the implications of households' ICT use for energy consumption (Røpke *et al.* 2010), we examined 48 activities organised into 10 groups: communication, entertainment, information, purchase and sale, work at home, education, hobbies and volunteer work, administration and finances, domestic work and management of the dwelling and, finally, health (Jensen *et al.* 2009; Røpke *et al.* 2010). Our study demonstrates how the generic functionalities of ICTs are integrated into all sorts of practices, and how many practices are transformed in the process, including practices with no obvious relation to the classical use of ICTs, such as different sports and do-it-yourself activities. For example, several informants went jogging and integrated ICT into this practice in various ways, such as using internet-based maps for planning routes, using running computers to measure the length and gradients of the route and to monitor their speed and pulse and video to record and analyse their performance on the computer afterwards. In addition, various net-based services helped in arranging running competitions with friends or strangers.

As this example illustrates, the integration of ICTs is often followed by a process of diversification as new features are added to a practice or a complex of practices. Similar dynamics have been noted by others; for example Karlsson and Törnqvist (2009) show how the practice of bird-watching has absorbed successive generations of ICTs to provide real time information on where to see rare birds with derived effects on transport. We cannot tell whether the ICT enabled variants of jogging, singing in a choir or bird-watching that we observed are widely diffused, or whether new practice entities have taken shape. Even so, our observations indicate important trends: although the use of ICT sometimes contributes towards reducing the energy-intensities of practices, the pervasiveness of ICT integration across everyday activities, and the related diversification of practices and complexes of practices, tends to increase energy-intensities.

ICT in relation to time and space

We now turn from the energy implications of the pervasiveness of ICT and the related diversification of practices to the temporal and spatial implications of integrating ICT in everyday practices and the consequences these have for energy consumption. This brings the second- and third-order effects mentioned in our introduction into focus. The growing use of ICT in relation to more and more activities – such as entertainment, reading the news, banking transactions and communication in general – supports a partial de-coupling of practices from their previous time-space location. One example of this is mobile broadband and smart phones with internet access, which make it possible to read the latest news or check the latest updates on Facebook while on the move. This partial de-coupling of practices enables new ways of weaving together practices across time and space, which contributes to a more fractured everyday timescape. This tendency is also supported by the increased 'digitalisation' of practices, which implies that the performance of an array of practices is 'gathered' in relation to the same devices, such as the smart phone or the laptop. As a result, the friction related to changing between different practices is reduced; in many cases, it literally only takes a mouse click or a finger sweep across the touch screen to shift from one ICT-supported activity to another, e.g. between chatting with a friend on Facebook and checking the latest news story.

Developments within ICT challenge previous notions of practices as time-and-place bounded entities. Despite bringing this aspect into view, we do not agree that ICTs have made details of time and place irrelevant for understanding human experience and social practices. As pointed out by others, such radical interpretations fail to recognise that institutional and collective time-structures (Pantzar and Shove 2010) as well as local pockets of order, such as the home (Ellegård and Vilhelmson 2004), still play a significant role in structuring, locating and coordinating individual activities and the interaction between people. Keeping these qualifications in mind, it is still obvious that profound changes have taken place within many areas of everyday life in parallel with the integration of ICT, for example in relation to mediated communication between family members (Christensen 2008, 2009), mediated peer group interaction of young people (Livingstone 2002) and coordination of face-to-face meetings through mobile phone-based micro-coordination (Ling 2004). In the rest of this chapter, we reflect on the implications of these kinds of changes for energy consumption.

Softening time constraints

The integration of ICT implies a softening of the time (and place) constraints of many practices that were previously limited (e.g. by opening hours). For example, bank transactions and communication with the authorities (e.g. in relation to income taxes) are increasingly done online via internet services. Similarly, online shopping can be done at any time. One of the most visible effects of the softening of the time (and space) constraints of practices is the use of ICT to 'activate dead time'. Small pockets of time not focused on one specific activity and often perceived as 'unproductive time', like waiting for the bus or commuting, have increasingly been filled with activities supported by mobile phones, laptops, tablet computers, e-book readers, etc. These activities include talking or texting with family members and friends, checking e-mails, playing computer games, listening to music or podcasts, checking the Facebook profiles of friends and watching movies or video clips, etc.

Filling time between other activities with media consumption is not an historically new phenomenon (e.g. Bull 2000). However, portable ICT devices, in combination with mobile broadband access, multiply the number of practices that can be engaged in while waiting or on the move. In a sense, the diffusion since the late 1990s of small and portable ICT devices (increasingly with mobile internet access) has contributed to a remarkable 'dispersion' in time and space of many practices, particularly practices related to entertainment and mediated communication.

Relations between practices – whether in the form of bundles or complexes – are supported by an ICT-based technological environment where smart phones and portable computers increase opportunities for multitasking, i.e. the simultaneous performance of two or more practices. Statistical evidence indicates that the combination of television viewing with other ICT-supported activities, such as communicating with friends or visiting websites, is widespread. ICT-supported activities seem to be especially flexible with regard to 'fitting in' with other practices, a feature that supports multitasking. In relation to the internet, Kenyon notes that:

> ...internet use may be expected to influence multitasking in two ways beyond the mere substitution of activities from offline to online: by increasing the number of activities that can be multitasked (activities *amenable to* multitasking); and by increasing the accessibility of a greater number of activities (activities *accessible for* multitasking).
>
> (Kenyon 2008: 291; author's own emphasis)

Like the internet, many uses of the mobile phone are capable of combining with other practices: for example, it is common to combine car driving with conversing on the mobile phone. The experience of one of the families included in our research gives a sense of how this works. This family consists of a couple living with their two sons in the countryside, about 100 km from their workplaces in

Copenhagen. As they spend several hours each day commuting by car, the parents have made it a daily routine to utilise their travel time for work- and family-related communication via the mobile phone. The father explains that in the morning he communicates with his work colleagues, and, on his return trip in the afternoon, he has long conversations with his wife or calls their sons, who are often alone at home after school.

The activation of 'dead time' and increased multitasking, enabled by the partial de-coupling of many practices from previous time and space constraints through the use of ICT, contributes towards a dense packing of everyday life. As the intersecting practices related to multitasking in general seem to relate to a diversity of projects – such as family, work and friendships – these seem to represent bundles of minimally related practices rather than complexes of mutually dependent practices. An example of this is the couple above, who combine commuting with communication related to both work and family. In energy terms, lives that are more densely packed are likely to produce second-order and third-order increases in the total consumption of energy, since the performance of each single practice involves energy use. In the concluding section, we return to this general trend, which counteracts the energy savings that may emerge in relation to specific practices.

Softening space constraints

As indicated above, the integration of ICT is often associated with a partial de-localisation of practices. An increasing number of practices can be performed without being in any specific place. However, it is obvious that there are significant differences between practices with regard to the potential for de-localisation. Some practices are more place-specific, and therefore constrained by space, than others: reading a bedtime story for a child, preparing dinner, do-it-yourself work in the home and the weekly bridge game in the club, are examples of practices that are difficult to de-localise.

The degree of place-constraint is determined by the social and material framework as well as the private frameworks of the individuals involved in the performance of the practice. More specifically, practices with a high degree of place-constraint typically (1) involve place-specific objects for their successful performance (e.g. the practice of preparing dinner), (2) require physical co-presence between two or more practitioners in order to be performed in a meaningful way and therefore imply coordination of the individuals' spatial movements (e.g. social gatherings with friends) and (3) involve normative expectations regarding the right place of performance. It seems that for many place-constrained practices, all these characteristics typically play a role. An example is preparing and eating the daily family dinner; besides involving place-specific objects like kitchen equipment, this is closely associated with the home and is imbued with a strong normative expectation of co-presence (the ideal of eating together).

The development of ICT, particularly the internet and mobile devices for communication and internet access, has nonetheless enabled the softening of space constraints for many practices. As the following examples illustrate, this happens in one of two main ways: first, by relaxing the constraints related to place-specific objects through easier access to information and services (e.g. work-related activities and entertainment); second, by easing the constraints related to the requirement of physical co-presence by increasing the possibilities of mediated interaction (e.g. Facebook, voice over IP, e-mail, etc.).

Working from home is one of the most visible examples of the de-localisation of practices through an easing of space-constraints. With the diffusion of desktop computers, laptops and internet broadband access, in combination with changes in the form and content of many work-related tasks that now involve the computer as the main 'work tool', it has become possible for many to work at home. The share of employees in Denmark working from their homes at least once a month increased from 21 per cent in 2000 to 29 per cent in 2009 (Statistics Denmark 2010). Simultaneously, it has become common for many to communicate with family and friends via mobile phone or the internet while at work. Modern ICT facilitates new patterns of communication across the traditional time-space divide between work and home, a feature described as 'blurring the boundaries' in the literature on ICT and the work/life balance (e.g. Gant and Kiesler 2002; Salaff 2002).

The increased numbers of employees working regularly from home should, in principle, result in a reduction in commuting. However, different dynamics may reduce this energy saving potential, as also pointed out by Black (2001) and Buliung (2011). For instance, our interviews show that the possibility of working from home for a few hours in the morning enables some employees to travel later, avoiding the heaviest traffic congestion on the roads. This flexibility makes it more attractive to travel by car instead of by train (Christensen 2008). The possibility of working from home can also make it more attractive to live further from the workplace than would be reasonable if it were necessary to commute to work every day (Jørgensen *et al.* 2006). These modifications in working and commuting represent negative third-order effects that partly offset the positive second-order effects of working from home.

ICTs also enable the de-localisation of the diverse practices involved in managing and maintaining friendships. As the following examples illustrate, friendships depend on a complex of interrelated and mutually dependent practices in which ICTs figure in different ways. First, ICTs (including social network services like Facebook) are used to maintain contact with friends and acquaintances who are no longer seen on a regular basis (e.g. due to moving to another part of the country or to a new job). Our 2007–2008 study showed that Facebook, which was still quite new at that time, was used to sustain a large network of friends and to resume contact with old friends and acquaintances (see Røpke *et al.* 2010). Our interviews include examples of informants using various internet-based services to stay in contact with old friends. For instance, one couple uses the web-based photo album Picasa to share photos with their former neighbours who moved to Singapore some years ago.

A second form of ICT-mediated interaction includes new internet services, which make it possible to establish new relationships and widen the 'social space' related to many practices. Well known and much studied examples of this are the use of internet dating services and virtual communities, such as online chatting (e.g. Hardey 2004; Carter 2005). The internet includes a diversity of possibilities for establishing new relationships, particularly in relation to shared interests. One of our interviewees was a 49-year-old male truck driver who is a dedicated user of the multiplayer online role-playing game World of Warcraft, and who develops sustained relationships with some of the fellow-players he meets through the game. Another example is a 28-year-old male doctoral student who is a member of a small association of composers. He and his fellow members have personal profiles on MySpace and other places, and they sometimes make contact with new people who share the same musical interests. These include international contacts, which in some cases are followed up by visits and face-to-face meetings.

These instances show how ICT, and especially internet-based services, have made it easier to establish and stay in contact with a larger number of friends and acquaintances via mediated and often asynchronous communication: (1) It is easier to reactivate old friendships; (2) It is easier to maintain peripheral acquaintances or friendships between persons who do not physically meet on a regular basis; and (3) new friendships can be established around a shared interest. Like a rolling snowball, many users of social networking services seem to build up larger and larger networks of social relations, many of which are quite peripheral.

Rather than reducing transport by replacing social interaction based on physical co-presence with mediated or 'virtual' interaction, and thereby contributing to a positive second-order effect, the new ICT-supported social networking possibilities might, in fact, imply more travel. As an individual's social network grows, the number of people they might travel to meet increases. More broadly, the use of internet-based services to maintain relations across large geographical distances may modify perceptions of distance. Thus, Urry writes: 'As virtual travel becomes an ordinary part of everyday life (...) it may transform what is experienced as near and far, present and absent.' (2004: 33). If the world is increasingly populated with geographically scattered 'known faces' – like old school friends, former neighbours or people with shared interests – this would seem to encourage more travelling (see also Carter 2005).

Handling increased complexity

As previously mentioned, the softening of the time-space constraints of many practices leads to an increasingly fractured timescape in individuals' everyday lives, with practices related to communication, work and leisure often intersecting.

Practices are packed closer together, and this often involves the need for multitasking. In general, ICT seems to support multitasking, for example, when adolescents simultaneously do homework, watch television, listen to music and text their friends via instant messaging. Also, independence and individual flexibility increases the need for continuous coordination of face-to-face meetings, which is again related to the softening of institutional and collective temporal rhythms.

With regard to complexity, ICT often plays a double role, which seems to involve a dialectic that induces further energy consumption. The integration of ICT into more and more practices contributes to the fragmentation of an increasingly complex everyday timescape. At the same time, ICT plays an important role in people's efforts to handle this complexity.

A 2010 advertisement campaign in the Danish media for the smart phone 'HTC Wildfire', illustrates this dialectic. The ad had the heading: 'All your friends and everything they are busy with gathered in one place.' The main message of the ad, which targeted young people, was that the HTC Wildfire smart phone makes it possible to gather information from several social networking services in one device. The caption below the picture of the phone and the corporate logos of Facebook, Twitter and Flickr etc. reads:

> You have lots of friends in a lot of places. Why not bring them closer to you? Get all your friends' updates on Facebook, Twitter and Flickr in one single feed. And when the friends call, you even get their latest status on Facebook.
>
> ('HTC Wildfire' advertisement 2010)

The main message of the advertisement is that the HTC Wildfire smart phone helps to gather the scattered digital information about friends in one place and at the 'right time'.

Finally, as already mentioned, mobile phones are used to coordinate the meeting of persons in time and space. Interestingly, this practice is closely related to change in the way we organise face-to-face meetings. Over the last two decades, making agreements based on the ideal of 'punctuality effected through clock time to a flexible and perpetual coordination effected through email and mobiles' (Larsen et al. 2008: 640), has been replaced by new forms of micro-coordination.

As these examples illustrate, ICTs enable a softening of practices' time and place constraints, and, simultaneously, promise to integrate data in ways that help people handle the complexity that follows. Using ICT to manage complexity can be regarded as a third-order effect related to the softening of time and place constraints.

Discussion and conclusion

In this chapter, we have not sought to assess the net energy impacts of the use of ICT, but rather to let everyday life take centre stage and point out effects that are not so visible in the approaches applied until now. Previous research has examined the impact of ICT on residential electricity consumption, but when it comes to second-order and third-order effects of ICT integration, the role of consumers has mainly been discussed in relation to a few selected applications, especially those in which using ICT is expected to reduce environmental problems. Such studies have focused on the 'dematerialisation' of practices, where 'virtual goods' replace material devices. For instance, life cycle assessments indicate a significant second-order energy saving potential related to purchasing music based on the downloading of MP3 files as compared to physical CD delivery (Weber *et al.* 2010). However, these savings might be partly outweighed if streaming or easy access to downloading low-price/free music results in a significant increase in the data traffic on the internet, thus resulting in increased energy consumption for the internet infrastructure.

Other studies have dealt with broader issues, like telework or teleshopping, where considerable second-order energy savings have been expected due to reduced travel. As already mentioned, the potential savings from telework might be counteracted by other changes in practice. Likewise, buying goods and services through the internet may reduce private trips to the shops, but as many goods are delivered individually by lorry or car, there is no necessary transport-related energy saving. Using the internet for services like bank transactions and communication with public authorities seems certain to reduce second-order travel but, in total, the impact of such changes may not be significant.

In contrast to those who focus on specific technologies and applications, our research examines the diffuse impacts of ICTs and their integration into multiple practices. We emphasise the many small changes taking place in relation to a large number of practices in which ICT plays a somewhat marginal role. The dynamics we describe emerge from the interplay between the 'logic' that is internal to a given practice, and seemingly external changes. Engaged practitioners are often interested in improving their performances – whether this implies finding new recipes for cooking or monitoring running results – and new technologies help achieve improvements. Many practitioners demonstrate considerable creativity in the integration of ICTs, and further develop the devices and services that are sold commercially.

When front-runner practitioners develop new ways of performing practices, the seeds are gradually diffused to others, and new practices-as-entities may result. In this process, new practices may contribute to the formation of new structural conditions in the form of physical and institutional patterns, as well as normative expectations – changes that sometimes force latecomers to follow suit. For instance, communication with the authorities increasingly requires access to the internet, and the coordination of activities and physical meetings between people depends, to a high degree, on the use of mobile phones.

In addition, we have argued that mobile technologies and the internet offer unique potential for softening the constraints of time and space. Many practices have become partially de-coupled from their previous time-space location, and a more fractured timescape has emerged. As mentioned above, this development increases opportunities for performing practices in new and interesting ways, for instance, by making it possible to cultivate extended social networks and engage in special-interest communities. However, the de-coupling also results in considerable challenges, such as those faced by people trying to combine and balance projects and practices that compete for time. This task is eased when ICT also allows people to carry out more practices within the temporal constraints of everyday life – for instance, when the friction between different practices is reduced, 'dead time' is activated, and multitasking is made easier. ICT can ease the coupling constraints that arise from the need to coordinate individuals' paths in time and space: with ICTs, many practices can now be performed in de-synchronised ways or without meeting in the same place. Cutting across time and space, ICT contributes both to the increasing complexity of everyday life and to the handling of this complexity – making possible the management of more practices within temporal constraints and the management of movements within a wider space.

Few of the many changes related to the integration of ICT in everyday practices are motivated by energy considerations. Still, some uses of ICT hold a potential for energy savings but, as our research shows, these are partially or fully counteracted by other ICT-supported changes in everyday practices that imply increases in energy consumption (including diversification of practices, cultivation of wider social networks, handling complexity, etc.). In detail, and in practice, the energy impacts of ICT depend on wider economic and political conditions, elaborated in Røpke (forthcoming). Seen from the demand side, people integrate ICT in practices when it is relevant and when they can afford to do so, and their ability to pay depends on both their income and the prices they have to pay for ICT equipment and services.

Our studies have been carried out during a period when many Danes experienced increasing incomes and rising wealth due to gains on property, reflected in increasing consumption. Simultaneously, the prices of ICT dropped significantly because of technological improvements (Moore's law) and, not least, access to cheap raw materials and cheap labour in sweatshops. Furthermore, it was not very important for people to be concerned about the energy-intensity of practices, because energy prices were modest in the period of study, and the availability of cheap transport did not encourage the use of ICT for replacing physical transport with 'virtual' travel and communication.

The supply side reinforces these trends, as businesses compete to fulfil the imagined wants of consumers, and as modest energy prices influence the selection environment for innovations, providing little incentive for designing energy-saving products and services. It is also important to recognise that there has been strong political interest in promoting ICTs, not for energy saving but for reasons of international competitiveness. Concerns about climate change and energy have entered the frame, resulting in efforts to strengthen public regulation of the energy use of consumer electronics. However, this control of the first-order effects is very far from catching up with the growing quantity of devices and related practice dynamics.

When it comes to second-order and third-order effects, ICTs have great potential for reducing the energy intensity of everyday life: the dematerialisation of various practices and reduced travel are promising prospects. But, as our research clearly demonstrates, the realisation of these potentials does not come about automatically, nor is it a simple effect of technological change. A protracted economic crisis or increasing energy prices may encourage the application of ICTs in ways that save energy. Public policies might actively encourage such trends. In addition to energy taxation, innovation policies can direct investments towards energy-saving applications. Similarly, efforts to reduce global inequality might increase the prices of raw materials and wages in sweatshops, making ICTs more expensive and perhaps limiting the range of practices with which they co-evolve.

Acknowledgements

We are grateful to Tomas Benzon, who helped to draw the figures, and to Elizabeth Shove, Nicola Spurling and two anonymous reviewers, who gave elaborate comments on a previous version of this chapter which has been published in Røpke, I. and Christensen, T.H. (Forthcoming) 'Energy impacts of ICT – Insights from an everyday life perspective', *Telematics & Informatics*, available online at: www.sciencedirect.com/science/article/pii/S07365 85312000184.

References

Adam, B. (1998) *Timescapes of Modernity: The Environment and Invisible Hazards*, Routledge: London.
Black, W.R. (2001) 'An unpopular essay on transportation', *Journal of Transport Geography*, 9: 1–11.
Buliung, R.N. (2011) 'Wired people in wired places: stories about machines and the geography of activity', *Annals of the Association of American Geographers*, 101(6): 1365–1381.
Bull, M. (2000) *Sounding out the City: Personal Stereos and the Management of Everyday Life*, Berg: New York.
Carter, D. (2005) 'Living in virtual communities: an ethnography of human relationships in cyberspace', *Information, Communication & Society*, 8: 148–167.
Christensen, T.H. (2008) Informations- og kommunikationsteknologi i familiens hverdag, Ph.D. dissertation, Department of Management Engineering, Technical University of Denmark: Kgs. Lyngby.
Christensen, T.H. (2009) '"Connected presence" in distributed family life', *New Media & Society*, 11: 433–451.
Christensen, T.H. and Røpke, I. (2010) 'Can practice theory inspire studies of ICTs in

everyday life?', in B. Bräuchler and J. Postill (eds), *Theorising Media and Practice*, Berghahn Books: New York.

Crosbie, T. (2008) 'Household energy consumption and consumer electronics: the case of television', *Energy Policy*, 36: 2191–2199.

Ellegård, K. and Vilhelmson, B. (2004) 'Home as a pocket of local order: everyday activities and the friction of distance', *Geografiska Annaler: Series B, Human Geography*, 86: 281–296.

Erdmann, L. and Hilty, L.M. (2010) 'Scenario analysis: exploring the macroeconomic impacts of information and communication technologies on greenhouse gas emissions', *Journal of Industrial Ecology*, 14: 826–843.

Freeman, C. (1992) *The Economics of Hope: Essays on Technical Change, Economic Growth and the Environment*, Pinter Publishers: London.

Gant, D. and Kiesler, S. (2002) 'Blurring the boundaries: cell phones, mobility, and the line between work and personal life', in B. Brown, N. Green and R. Harper (eds), *Wireless World: Social and Interactional Aspects of the Mobile Age*, Springer Verlag: London.

Hägerstrand, T. (1985) 'Time-geography: focus on the corporeality of man, society, and environment', in S. Aida (ed.), *The Science and Praxis of Complexity*, United Nations University, Tokyo.

Hardey, M. (2004) 'Mediated relationships: authenticity and the possibility of romance', *Information, Communication & Society*, 7: 207–222.

Hilty, L.M. (2008) *Information Technology and Sustainability: Essays on the Relationship Between ICT and Sustainable Development*, Books on Demand GmbH: Norderstedt.

IEA (2009) 'Gadgets and gigawatts: policies for energy efficient electronics', International Energy Agency: Paris.

Jensen, J.O., Gram-Hanssen, K., Røpke, I. and Christensen, T.H. (2009) 'Households' use of information and communication technologies –a future challenge for energy savings?', Proceedings of ECEEE Summer Study, Cote d'Azur: France.

Jørgensen, M.S., Andersen, M.M., Hansen, A., Wenzel, H., Pedersen, T.T., Jørgensen, U., Falch, M., Rasmussen, B., Olsen, S.I. and Willum, O. (2006) 'Green technology foresight about environmentally friendly products and materials –the challenges from nanotechnology, biotechnology and ICT', Danish Ministry of the Environment: Copenhagen.

Karlsson, K. and Törnqvist, E.K. (2009) 'Energitjuv eller sparverktyg? Om användning av informations – och kommunikationsteknologi i hushåll', Elforsk rapport 09:86.

Kenyon, S. (2008) 'Internet use and time use: the importance of multitasking', *Time & Society*, 17: 283–318.

Kuehr, R. and Williams, E. (eds) (2003) *Computers and the Environment: Understanding and Managing their Impacts*, Kluwer Academic Publishers and United Nations University: Dordrecht.

Larsen, J., Urry, J. and Axhausen, K. (2008) 'Coordinating face-to-face meetings in mobile network societies', *Information, Communication & Society*, 11: 640–658.

Ling, R. (2004) *The Mobile Connection: The Cell Phone's Impact on Society*, Elsevier Inc. and Morgan Kaufmann Publishers: San Francisco.

Livingstone, S. (2002) *Young People and New Media*, Sage Publications: London.

Meier, A. (2005) 'Standby: where are we now?', Proceedings of ECEEE 2005 Summer Study.

OECD, (2010) 'Greener and smarter: ICTs, the environment and climate change', OECD: Paris.

Pantzar, M. and Shove, E. (2010) 'Temporal rhythms as outcomes of social practices: a speculative discussion', *Ethnologia Europaea*, 40:19–29.

Pred, A. (1981) 'Social reproduction and the time-geography of everyday life', *Geografiska Annaler*, 63 B: 5–22.

Reckwitz, A. (2002) 'Toward a theory of social practices: a development in culturalist theorizing', *European Journal of Social Theory*, 5: 243–263.

Røpke, I. (2009) 'Theories of practice: new inspiration for ecological economic studies on consumption', *Ecological Economics*, 68: 2490–2497.

Røpke, I. (Forthcoming) 'The unsustainable directionality of innovation – the example of the broadband transition', *Research Policy*, available online at: www.sciencedirect.com/science/article/pii/S0048733312001011.

Røpke, I., Christensen, T.H. and Jensen, J.O. (2010) 'Information and communication technologies: a new round of household electrification', *Energy Policy*, 38: 1764–1773.

Røpke, I., Gram-Hanssen, K. and Jensen, J.O. (2010) 'Households' ICT use in an energy perspective', in J. Gebhardt, H. Greif, L. Raycheva and C. Lobet-Maris (eds), *Experiencing Broadband Society*, Peter Lang: Frankfurt am Main.

Salaff, J.W. (2002) 'Where home is the office: the new form of flexible work', in B. Wellman and C. Haythornthwaite (eds), *The Internet in Everyday Life*, Blackwell Publishers: Oxford.

Sandberg, E. (1993) 'Electronic home equipment Leaking electricity', Proceedings of ECEEE Summer Study: The Energy Efficiency Challenge for Europe.

Schatzki, T.R. (1996) *Social Practices: A Wittgensteinian Approach to Human Activity and the Social*, Cambridge University Press: Cambridge.

Schatzki, T.R. (2002) *The Site of the Social: A Philosophical Account of the Constitution of Social Life and Change*. The Pennsylvania State University Press: Pennsylvania.

Schatzki, T.R., Knorr-Cetina, K. and von Savigny, E. (eds) (2001) *The Practice Turn in Contemporary Theory*, Routledge: London.

Shove, E. and Pantzar, M. (2005) 'Consumers, producers and practices: understanding the invention and reinvention of Nordic walking', *Journal of Consumer Culture*, 5: 43–64.

Shove, E., Trentmann, F. and Wilk, R. (eds) (2009) *Time, Consumption and Everyday Life: Practice, Materiality and Culture*, Berg: Oxford and New York.

Shove, E., Watson, M., Hand, M. and Ingram, J. (2007) *The Design of Everyday Life*, Berg: Oxford.

Shove, E. and Pantzar, M. (2005) 'Consumers, producers and practices: Understanding the invention and reinvention of Nordic walking', *Journal of Consumer Culture*, 5: 43–64.

Southerton, D. (2009) 'Re-ordering temporal rhythms: coordinating daily practices in the UK in 1937 and 2000', in E. Shove, F. Trentmann and R. Wilk (eds), *Time, Consumption and Everyday Life: Practice, Materiality and Culture*, Berg: Oxford.

Statistics Denmark (2010) Flere arbejder hjemme. [Homepage of Statistics Denmark], available online at: www.dst.dk/Statistik/BagTal/2010/2010–02–24-Hjemmearbejde.aspx.

Steinmueller, W.E. (2007) 'The economics of ICTs: building blocks and implications' in R. Mansell, C. Avgerou, D. Quah and R. Silverstone (eds), *The Oxford Handbook of Information and Communication Technologies*, Oxford University Press: Oxford.

The Climate Group (2008) 'SMART 2020: enabling the low carbon economy in the information age', Global eSustainability Initiative (GeSI).

Urry, J. (2004) 'Connections', *Environment and Planning D: Society and Space*, 22: 27–37.

Warde, A. (2005) 'Consumption and theories of practice', *Journal of Consumer Culture*, 5: 131–153.

Weber, C.L., Koomey, J.G. and Matthews, H.S. (2010) 'The energy and climate change implications of different music delivery methods', *Journal of Industrial Ecology*, 14: 754–769.

Wilk, R. (2009) 'The edge of agency: routines, habits and volition', in E. Shove, F. Trentmann and R. Wilk (eds), *Time, Consumption and Everyday Life: Practice, Materiality and Culture*, Berg: Oxford.

Willum, O. (2008) 'Residential ICT related energy consumption which is not registered at the electric meters in the residences', Willum Consult and DTU Management Engineering: Copenhagen.

Yi, L. and Thomas, H.R. (2007) 'A review of research on the environmental impact of e-business and ICT', *Environment International*, 33: 841–849.

5 Mundane materials at work
Paper in practice

Sari Yli-Kauhaluoma, Mika Pantzar and Sammy Toyoki

Introduction

Environmental debate has suggested that a less paper intensive office is an important step towards sustainability. Based on a purely technological, product- or material-based perspective, this debate has been hindered by the misguided notion that good matter ("dematerial") always replaces bad matter rather than coexisting alongside it, and that change is always revolutionary (Duguid 1996). Indeed, both technologically-based (deterministic) visions of dematerialization and those stressing the roles of "green office" administrators, fail to understand entire and changing infrastructures of competence, desires, emotions, and technologies that support and live with existing practice constellations. Sellen and Harper (2003) have pointed out that some reading activities require laying paper out, for example, on a table. Flexible navigation and manipulation of paper is essential in order to read and write across documents. New ways of working need to take old ways into account. In the following, we suggest that the role of paper is central to these old ways. "If organizations can understand where paper is useful and, more important, why it is useful, they may be able to make better-informed judgements about how new and old technologies will work together" (Sellen and Harper 2003: 19).

Knowledge is not an external, enduring, or essential substance but a dynamic and ongoing social accomplishment. As Orlikowski suggests, a practice view of knowledge leads us to understanding:

> knowing as emergent (arising from everyday activities, and thus always in the making'), embodied (as evident in such notions as tacit knowing and experiental learning), and embedded (grounded in the situated sociohistoric contexts of our lives and work).
>
> (Orlikowski 2006: 460)

To this list, Orlikowski adds another important dimension on which this chapter also focuses, namely, *knowing is also material*. Informed by this perspective, we examine the detailed material anchoring of paper-related routines and practices as they occur in a university back-office (for details, see the methodological notes included at the end of this chapter).Through this research, we highlight how digitalization is challenged by the "work" that the mundane material of paper does in practice, showing how these two systems (paper-based and digitalized) have in fact co-evolved.

Our main suggestion is simple: different forms of paper in an office, from Post-it™ notes to flip-charts and books, bridge between work practices in time and place(s). In such an environ, work practices, with their hard-to-detect infrastructures, flows of discursive knowledge, and teleoaffective framing, have, over a long period of time, evolved within a paper-based system. Indeed, it is our contention that administrative practices have, in large part, been shaped by the materials through and around which they have developed. We argue that it is impossible to understand change brought by digitalization in office contexts if current paper-related practices are poorly understood (Duguid 2005; Levy 2001; Sellen and Harper 2003). The entry of new material (here "dematerial") in a life of a bureaucracy can only be comprehended if the actual constellations of current (or transforming) practices within offices are thoroughly grasped, making the co-existence and cross-fertilization of new and old practices the most obvious path of inquiry.

Our chapter complicates the picture further by showing that change does not happen in a linear way, and that dematerialization (i.e., the absolute or relative reduction in the quantity of materials required to serve economic functions) in one place could lead to re-materialization in another place. Our main interest is in specifying the way materiality, here paper, structures and enables existing practices within an office. The line of thinking we are adhering to draws reference to such concepts as "performance" (Trentmann 2009), "affordance" (Sellen and Harper 2003) and "functionality" (Duguid 2005) of material. Our first tentative proposition is that in the practices included in our study much (social order) remains the same even if digitalization takes place. Here, elements of work practices become "fossilized" (Shove and Pantzar 2006), as their continuous existence is based more on some sort of ritualized behavior than direct functionality. Our second proposition is in slight contradiction to the first: in some cases, due to already existing elements and ecosystems, change can be fast once certain thresholds have been crossed. However, in this case the transformation is not necessarily recognized as a being radical, as will be suggested in the "Discussion" part of this chapter.

The field work took place in the administrative department of personnel and legal affairs in our own institute, the former Helsinki School of Economics (from the beginning of 2010 known as Aalto University, School of Economics). We conducted 14 in-depth interviews in total during five months from September 2009 to January 2010. The interviews lasted between one and three hours. All interviews were taped and most were transcribed verbatim. A grounded theory analysis of the data was undertaken. While our approach to data analysis was inductive and rigorous, we recognize also that it reflected our idiosyncratic understandings regarding what constitutes interesting first order data from which to build theory. Below we condense some of the most relevant findings.

Work in offices – some preliminary findings about paper-related practices

Our ethnographic study focused on a mundane material at work, namely paper. The aim was to find out how office workers explain their everyday use of paper. Originally, we anticipated that paper-related practices would be linked to rather general categories such as "printing," "copying," "writing", or "reading." To our surprise, however, the paper-related activities people actually talked about were significantly more nuanced and creative. Our interviewees also felt a strong need to express resistance towards reliance on digital documentation. Table 5.1 summarizes the various types of practices to which paper is tied in office work.

In the following, we briefly describe the paper-related practices we found. We deliberately emphasize the ways in which the paper format permits social and temporal bridging between various work practices.

Social coordination

Administrators take, send, give, and receive various types of printouts or forms in a paper format. In our study, we noticed that administrators printed out pieces of paper (that existed as digitalized documents) and used these as "boundary objects" in problem-solving situations, for example, when they needed to ask or to present somebody with something.[1] The documents that our interviewees took and held in their hands and used as points of reference for inquiries or discussions included bills, parts of larger documents, or some kinds of reports. A project manager in an administrative office gives an example of how paper is used as a boundary object and explains its advantages over, for instance, computer or information technologies in situations requiring detailed discussion or enquiry:

> If I would show [our administrative manager] the screen of our [electronic] system, point out to a specific case, and ask about her opinion, she would not understand what we are talking about.... But, if I go to her and show her a piece of paper containing, for example, a motion from a university department ... and say that here are the details, then she can make a comment on it.... I could send the same thing by e-mail but in those cases I would need to explain it to her anyway.

Indeed, it has been well established that information exchange is one of the most common office practices in which paper is used (Carlile 2002). Paper-based practices foster synchronic order by facilitating communicative bridging between actors. How such practices also create order by bridging between the past and the future is a less familiar topic in academic studies of administration. This is discussed next.

Table 5.1 Seven essential paper practices in offices

	Key action to do with paper/on paper	Type of paper	Essential dimensions of paper as material	The role of paper	Time of existence of the paper	Locus of paper	Examples of phrases said about paper
Social coordination (bridging) practices	To transfer knowledge	Printouts, forms	Mobility, convenience	Paper as a boundary object	Short: minutes, hours or some days	In hand, in briefcases	"If I have a question for somebody, it is easier to print out the invoice and take it with me than to ask the person to come here or something else."
Remembering (reminding) practices	To remember to do a task	Post-it™ notes, pieces or sheets of paper, printouts	Visibility, simultaneity	Paper as a reminder	Short: minutes, hours, or some days	Office tables, computer screens	"It is much better to have it on paper, because it then stays here. [Otherwise] it might easily get lost in my e-mail box."
Anticipatory practices	To get ready for a particular task in the very near future	Sheets of printouts stacking up into piles	Manageability, confidentiality	Paper as an information carrier	Medium: weeks or months	Office tables, office cabinets' printers, copying machines	"There is a tight deadline when the agenda needs to be finalized. As soon as I can get prepared, I will start printing out the attachments."

Sketching practices	To think and to create something new	White or blank pieces of paper	Flexibility, accessibility	Paper as a thinking device	Short: hours, or some days	Flip charts, recycling bins, digital archives	"It sort of makes things simple, the paper and pen, in the way that you focus on the subject. Thinking requires space. But a computer kind of takes you over and constrains you."
Modeling practices	To gain ideas about *how* a job has been done in the past	Printouts, books	Accessibility, readability, specificity, confidentiality, appropriate similarity	Paper as a guideline	Long: from months to years	Binders, folders, office tables	"For someone who cannot [do something] or who does not know the case at all it would be easier to follow [instructions] on paper than to have two windows open on the computer screen."
Verifying practices	To get prepared for questions and problems	Original documents, Copies	Conclusiveness, immutability	Paper as a piece of evidence	Medium: weeks or months	Beside electronic systems, personal files	"…mistakes always happen and then it is good to have some kind of material that you can make comparisons with."
Back-up practices	To hold knowledge on the current state of affairs just in case	Documents that have already been processed	Immutability	Paper as staging post	Long: from months to years	Office tables, personal files	"I do not keep them forever, perhaps a year or two, because they might contain matters that pop up again."

Remembering practices

Most of the offices we visited were more or less full of paper. Administrators actively built up different kinds of paper arrangements on or near their office tables. There were colorful Post-it™ notes, separate sheets of paper as well as paper stacked up in piles. Our interviewees repeatedly talked about the importance of such arrangements in reminding people about specific tasks and things to do.[2] For instance, the Post-it™ notes that we saw were often placed either directly on computer screens or at least somewhere near the computer or telephone. A managing director explains the advantage of Post-it™ notes in addition to using the computer:

> Even though I normally use the computer, here I have Post-it™ notes to write on, when I talk on the phone.... The Post-it™ notes..., for example, in the morning, I may have begun a conversation on the phone already in the car, and while I start-up my laptop here, I may make some notes on the [conversation]. It is because they are right away available....Then I also have here two or three Post-it™ notes that I wrote in the car while waiting in the traffic lights. I just couldn't write my [notes] down anywhere else. These are now waiting for me to take care of them, that's their purpose.

Here, Post-it™ notes extend the use of a technology by being ready-at-hand *while* setting up one's computer, talking on the phone, or driving the car. These slips of paper help to bridge the gap between times and places when the use of a more complex technology is not possible, serving as memory-aids or cues for future action.

Anticipatory practices

Administrators make and use piles of paper to remind them of different matters amidst the hectic rhythm of daily life. However, this is not the only reason for building up such stacks. Our study shows that another important reason is that administrators have to prepare for specific tasks, many of which have critical deadlines. In addition to representing the sequence in which tasks need to be taken care of, thus structuring the work process, paper piles also have a motivating feature. Once they reach a certain height, piles induce a sense of urgency – providing a tangible representation of a "build-up" that needs clearing. Computers do not reproduce this kind of physical build-up.

An example of this kind of task/anticipatory practice can be found when administrators prepare an agenda for a meeting. Previous research has already indicated that agendas are important tools in planning, preparing for, and structuring meetings (Volkema and Niederman 1996). In practice, preparing an agenda, and gathering and compiling all the documents involved, may constitute an enormous job with many phases.[3] For instance, the personnel and legal affairs secretary in the department of the university that we studied, explains that preparing an agenda often begins by first contacting the relevant officials and then mapping out matters that might be included in it. From an administrative perspective, the most critical preparatory task after that of determining the content is to receive and actively collect supporting material, meaning various kinds of documents and attachments usually provided by the officials involved. These documents are then carefully sorted and put in an order that corresponds to the agenda of the forthcoming meeting. The secretary in our study describes the key challenges associated with the work of producing an agenda and assembling all the relevant paperwork:

> There is a tight deadline when the agenda needs to be finalized. The printing alone takes a lot of time when you need to print out tens of pages of paper.... As soon as I can get prepared, I will start printing out the attachments, or kind of making copies of them.... In principle, the deadline is on Monday at noon. That is when I need to have the materials to get it ready for distribution during the same day. ... Only once have I been late. Then it went for distribution on the next day. I did not receive the attachments until 1 p.m. Then we started making copies of them here and it went a bit tight. But sometimes there are just matters that take so much time that you never get them on time. Unfortunately.

Sketching practices

In addition to printouts and other documents lying on desks "on hold," we saw various forms of paper ready-at-hand in the form of flip-charts, paper taped on office walls, or scattered on office surfaces and abandoned in recycling bins. What these seemingly different kinds of paper held in common, was their function as a "sketching" material. Sketching often involves writing, and most importantly, conceptual designing, e.g. drawing (Henderson 1991; Ewenstein and Whyte 2009). Usually, we associate drawing with the work of architects (Schon and Wiggins 1992), designers (Michlewski 2008), and engineers (Bechky 2003). We found that people in offices make drawings too, grabbing at sketch-papers and working on them when complex matters require elaborating and processing. A fundamental feature of sketches is that they are typically part of work-in-progress, meaning that there is usually an element of incompleteness about them (Ewenstein and Whyte 2009). Sketches reflect and enable processes of thinking and communication (Henderson 1991). As the manager of an administrative service center puts it:

> I think, it is so much better to describe the logic of thinking on paper and even draw something about it at the same time. It is the drawing in our job that often opens up the logic, how things go.

Modeling practices

In the offices that we visited, we saw different kinds of binders and folders (more or less full of printouts), piles of thick documents with attachments, and different kinds of administrative and professional note pads. An essential feature, connecting all of these various types of paper, was that they were used to contain and share detailed information and knowledge about how to carry out the various tasks at hand. As with other occupations, novices and newcomers to office work face the basic question of "how to do the job?" (Brown and Duguid 1991). Finding out is a matter of shared participation in the practice (Lave and Wenger 1991), but instructions remain important.[4]

The role of written instructions alongside watching, listening, and inquiring has been acknowledged by those who study processes of learning (Pinch et al. 1997). Recently, however, their significance has been downplayed as learning by instruction is deemed to be a top-down process and not one based on interaction. As the requirement for employees to take initiative and make independent contributions increases, the relevance of learning by instruction is perhaps diminishing (Håland and Tjora 2006). Nevertheless, our study provides some evidence to the contrary, demonstrating that in some areas of administrative life, written instructions are still extremely relevant. A secretary in the department of personnel and legal affairs explains:

> As I started in this job, I first wrote some instructions down [for myself] since I did not want to ask [somebody else] every time [I wanted to do something]...

While instructing may be a top-down process, using and taking note of instructions is an interactional practice involving intra- and interpersonal reflection (Mead 1934). As such, it should be viewed as a fundamental method of learning and managing the flow of office work. The citation below sheds some light on why instructions in paper form have, very probably, endured to this day:

> The department of personnel and legal affairs has a folder [with restricted access] in the university intranet. But, I think that it would really be easier to use [paper] if you would need to do something. For someone who cannot [do something] or who does not know the case at all it would be easier to follow [instructions] on paper than to have two windows open on the computer screen, one being the vineyard [i.e. the computer software] and the other the instructions for using that software.

We propose two key reasons for use of written (paper) instructions. First, written instructions are used, made sense of and understood by the individual her/himself as s/he seeks to apply them to her work. Also, the individual is often alone when s/he has to put the instructions into action. This is often why workplace tools and information technologies are developed. Second, written instructions are frequently used in an office context to operate some kind of computer software program. What, then, is the most practical way to read and follow such instructions? One possibility is to switch constantly between different computer screens. Another is to work with two wholly different media, such as paper and a computer, making the process perhaps more familiar and perhaps easier on the eyes. Ultimately, there are different ways of using and engaging with written instructions.

Verifying and backing-up

In administrative offices it is important to do things right, and various techniques have been developed to help office staff complete tasks on time and according "to the book" (Rämö 2002). Interestingly, our study shows that administrators do not only depend on existing processes: they also anticipate and prepare for encountering and tackling mistakes and problems. On the one hand, they do this by holding on to original documents in paper format, even when electronic versions also exist. On the other hand, they make copies of those documents that they need to take administrative work processes forward. These "verifying practices" are crucial in ensuring and proving that things have been done, and in checking that they have been done correctly:

> In the ideal world, it is enough when the information is in an [electronic] system and then nobody needs to check it any more. But, mistakes always happen and then it is good to have some kind of material that you can compare. [You get to see] whether things went properly. Therefore you still need those papers.

Office workers constantly struggle with the dilemma of whether to save and keep specific documents or throw them away.[5] People often hang on to documents as a means of hanging on to knowledge about the current state of affairs *just in case*. It is difficult to throw paper documents away because some may contain information that *will* be important in the future. We refer to this kind of saving and storing as a "back-up practice" in office work:

> I do sometimes take a look at in what meeting was it when we talked about that [matter] and what did we say about this matter and what kind of notes did I make in that paper. I still have them [the proceedings] in that pile.... In these office cabinets I even have older [proceedings].... I do not keep them forever, but perhaps a year or two, because they might contain matters that pop up again.

Material ordering of work practices

The examples above suggest that we need to learn more about the deep-seated material practices of office life before speculating about the prospect of dematerialization. When analyzing the challenges related to, say, filing and disposing of documents, we need to understand the material basis through and around which such practices have developed. For instance, different arrangements of paper (e.g., paper stacks) in specific locations in an office suggests a spatio-temporal ordering of working practices. Some paper arrangements are kept close to the center of action, maybe next to the computer or waste basket: these relate to tasks to be addressed immediately and in a certain order. Other stacks, including those stored in more peripheral parts of the office, relate to jobs and problems to be worked on at a later date. It is useful to focus on paper-based practices, but if we are to understand how they develop and change, we need to show how they relate to computerized practices. For instance, it would be safe to say that similar forms of spatio-temporal ordering go on in the digital realm. Digital files, documents, and folders on a computer "desktop" can be arranged and grouped (if not stacked) in ways that resemble paper-based forms of material ordering: digital items are also "thrown" into the "recycle bin" or "dragged" from one place to another to mark priority or urgency. Similarly, a full email inbox can make people as anxious as they would be when facing a more visible backlog of paper. As unremarkable as these observations may be, they suggest a link between material and digital practices. This link could be characterized as a kind of epistemic spillage from one realm to another. Individuals have some sense of continuity in that certain habits, rooted in paper-related practices, transfer across to the digital context. This suggests that the various spatial and temporal orders that paper arrangements create in an office space, and that are sometimes extended into the digital realm, relate to underlying forms of narrative sense-making (Bruner 1986). Rather than confronting a bustling chaotic flow of indistinct and unstructured elements, people encounter a world that seems to make sense because their consciousness interprets and gives meaning to these elements, identifying them as parts of a larger whole (Polkinghorne 1988). People often describe the arrangements they find in others' offices (or the icon and folder arrangements on computer desktops) as "organized chaos." However, what looks like a disorderly mess to the outside observer may nonetheless make perfect sense to the coordinator who has built, and who can therefore see, comprehensible networks of meaning:

> I am pretty good with piles. I know roughly what is in different piles and if someone comes and asks me [about a piece of paper], I can say that okay, it is here. On the other hand, if someone else comes to take care of my job [in my office] she does not necessarily find it at all.

By and large, it is our contention that the paperless office will not take hold in the near future unless paperless systems provide a high level of continuity with existing, *embodied* office practices. This is unlikely to happen in that digital approximations of material ordering cannot supplant physical doing and organizing.

Dematerializing the back-office – emerging themes

Our data suggest there are various reasons why paper is used in office life and provides some clues as to how these material activities unfold. According to our findings, people use paper: to transfer knowledge (social coordination), to remember to do some specific task (remembering practices), to think of and to create something new (sketching practices), to find out how a job has been done in the past (modeling practices), to prepare for tasks arising in the very near future (anticipatory practices), to cope with possible questions and problems (verifying practices), and to hold onto knowledge about the current state of affairs (back-up practices). What do these findings imply for an assessment of the potential for the dematerialization and digitalization of office work? What follows is a list of emerging themes for further research.

Offices as social-material complexes

Researchers have a tendency to underestimate the need to understand the broader setting of material/digital culture, something Williams (1974) describes as the "social-material complex." Duguid puts it this way: "Like an exasperated gardener, we snip triumphantly at the exposed plant, forgetting how extensive established roots can be" (Duguid 1996: 2).

Many seemingly trivial aspects of mundane life are deeply rooted in grand systems of representations, skills, and materials that have evolved historically (Levy 2001; Sellen and Harper 2003). For instance, the use of Post-it™ notes and flip-charts has developed alongside individual and organizational working practices (Yates and Orlikowski 2007). We might therefore argue that the administrators' conservatism and their reluctance to change is not simply a matter of irrational resistance. Rather, it is just as likely to be based on the failure of concepts and accompanying technologies to meet important needs and contextual requirements that are themselves anchored in past and present systems of practice.

Accordingly, we suggest that to understand the digitalization of office practices, we should focus on the emergence of new performative configurations consisting of changing material conditions (objects, technologies, and material culture in general), competences (human bodily activities and embodied skills) and images (meanings and representations of existing and emerging practices) (Magaudda 2011; Pantzar and Shove 2010). At the same time, it is important to recognize that because office life is already populated by the many elements of which current practices are composed, and because such practices constitute a larger system of practices, movement towards radically dematerialized forms of office life is slower than many proponents of the digital revolution would like to admit (Negroponte 1995; Toffler 1970). The more likely scenario is one in which old-fashioned, paper-related practices, and modern ones based on digital technology, will *co-evolve*.

Co-evolving constellations of practice

In the university that we studied, several different electronic systems link in to the management and administration of various types of monetary transaction within the institution. For example, the electronic system of personnel management links to the electronic system used for paying wages and other remunerations. From an administrative perspective, these are not easy systems to manage. While some university staff have permanent contracts, many others work on a temporary or short-term basis (Lamond 1997; Conley 2008). When temporary work contracts come to an end they are either renewed or terminated. This creates additional work for the administration and means that the electronic wage system has to be constantly updated. The challenge of keeping both electronic systems continuously updated has led to *the coexistence of paper- based and digital systems*. Accordingly, one paper copy of an original employment contract is always given to an employee, while the other original paper copy is filed by an administrator. At the same time, all key information in the original contract is typed into the electronic system (in personnel management) so that correct payments will be made from the wage system. In addition, a paper copy of the original contract is sent by courier to the administrative service center that verifies payment.

The reason for the coexistence of paper and electronic systems is that administrators need to be prepared for problems and mistakes, either in the use of information technology or in the administrative processes themselves. Previous studies have already pointed out that the detection and correction of errors is not necessarily easy in complex organizations (Edmondson 2004; Dholakia *et al.* 2009). In this study, we found that paper has an important function in helping to detect and correct mistakes in electronic systems, which may have emerged, for instance, through simple typing errors. It is for just such purposes that electronic systems are always accompanied by "paper trails."[6]

Final words

The fact that administrators have piles of paper in their offices is not the result of some random process. These stacks of paper link past decisions and operations through into current working practice. They also embody a forward-looking perspective. Administrators accumulate piles of paper as they prepare for upcoming tasks. A university administration office (or any kind of office) could thus be likened to a complex rhythm-based organism (Lefebvre 1991, 2004), representing a kind of junction box in which past and forthcoming practices intermingle.

In our view, different kinds of paper and different forms of paper arrangements affect the synchronic and diachronic ordering of office work, though it was the latter that our interviewees stressed the most. More generally, our findings suggest that the route towards the paperless office is likely to be longer and more winding than first anticipated by Alvin Toffler and other futurists in the 1970s.

To be sure, the office of the future will be a very different place compared to the office of today. New generation mobile workers, novel web-based applications and devices, emerging document exchange standards, and better scanning and digital archiving tools all herald fundamental changes in office life. Increases in the cost of natural resources, including paper and energy, will certainly affect the way that office work develops. Our message nevertheless remains the same: the success and eventual institutionalization of future (digital) practices depends on there being sufficient continuity between old paper-based practices and whatever new office practices might emerge.

Methodological notes

Our research approach resembles that of "at-home ethnography." "At-home ethnography" has two key features. The first is that there is "natural access" (Alvesson 2009). For us, "natural access" means that we operated in a context with which we are already familiar. We all have several years of experience as researchers in our institute. During these years, we have developed somewhat detailed knowledge of various issues relevant for this study, such as knowledge of the administrative offices in the university and of administrative meetings regularly taking place at our institute. Beside access, the other important element in "at-home ethnography" is that a researcher is an active participant in the cultural setting that s/he studies (Alvesson 2009). As researchers at the Aalto University, School of Economics we are not administrators, but as employees in the institute we confront and constantly participate in many kinds of administrative practices. We are in close contact and communication with administrative personnel in our own department and in some cases with the department of personnel and legal affairs. For these reasons it seems reasonable to claim that our method is one of "at-home ethnography."

We adopted a combination of methods, including fieldwork methods of ethnography, such as conversing, observing, working, and photographing, using these techniques to collect different kinds of empirical data (Ybema *et al.* 2009). We followed Nicolini's suggestion to focus on the "sayings and doings ... and the active contribution of artefacts [to constitute] a useful and practical starting point to orient the ethnographic gaze towards practices" (Nicolini 2009: 123). To be precise, we conducted 14 in-depth interviews between September 2009 and January 2010. These lasted between one and three hours, all were recorded and most were transcribed verbatim. Six were with administrators in the department of personnel and legal affairs in the Aalto University, School of Economics, and three with people working in the administrative Service Center, Pave. Aalto University has outsourced many of its administrative tasks to the Service Center and there is close contact between administrators in both organizations. We also interviewed two managers from large private companies and three independent professionals: a journalist, a consultant, and an architect. We made these interviews in order to compare paper practices in different working environments and reveal and clarify paper practices that were specific to administrative functions (Nicolini 2009: 132).

Questions focused on the interviewee's work and more specifically on the kinds of paper involved, and on how and why paper is used. The interviews were structured as discussions, which meant that we listened carefully and followed up with further questions (Forscy 2010). Finally, in addition to in-depth interviews, informal discussions, observations on administrative work, and participation as a university employee in administrative activities, we took photographs.

More particularly, we photographed the offices of all our respondents, using these to show how much paper there was in use, where and how it was located, and how it related to other objects and materials in the office setting (Harper 2003).

Notes

1. Boundary objects are "objects which are both plastic enough to adapt to local needs and the constraints of the several parties employing them, yet robust enough to maintain a common identity across sites" (Star and Griesemer 1989: 393). Boundary objects in organizational settings in particular are seen to transform knowledge, which is argued to be particularly relevant for the purpose of creating "common ground in order jointly to solve problems" (Zeiss and Groenewegen 2009: 90).
2. This is an important point to notice, particularly as existing research demonstrated that working life can be hectic (Kuhn 2006), disrupted (Staudenmayer *et al.* 2002) and fragmented, as several separate (and sometimes even independent) entities are intertwined in complex ways (Macpherson and Clark 2009). In this kind of working context the challenge is to capture and hold on to separate tasks and not to forget or lose sight of them.
3. For instance, the issue of confidentiality is particularly important in that it is directly related to the question of data or information sharing (Jones 2003) and therefore has concrete consequences for paper practices in offices. For example, in learning about the task of preparing agendas for meetings we discovered that supporting documents often contain confidential information. Moreover, confidentiality is one of the main reasons why complete agendas (including attachments) are distributed in paper format rather than via the Intranet.
4. Participation in practice can take many forms, which means that different kinds of doings may advance learning depending, for example, on the type of job or the person who is supposed to learn. So far, watching (Gherardi and Nicolini 2002) has been recognized as one important way of learning the essentials in various different kinds of occupations or tasks. Lave and Wenger suggest that as newcomers strive to learn a job, they often pay close attention to the practice (Lave and Wenger 1991) also observing how more senior professionals do their work (DeFillippi and Arthur 1998; Prentice 2007). In addition listening (Cooper 1997; Jacobs and Coghlan 2005), and enquiring (Patriotta 2003) are fundamental aspects of learning on the job.
5. So far, issues of organizational sustainability have focused on waste paper (Porter and Córdoba 2009), but processes of discarding are more complex, and are related to wider discussions of knowledge management (Hutchinson and Quintas 2008).
6. Another example of the coexistence of paper and electronic systems concerns billing. In the university that we studied the aim is to receive as many incoming invoices as possible in electronic form. These electronic bills then circulate in the system, passing from one person to another. However, some invoices still arrive in paper format. This may happen when bills are sent from abroad or when private persons submit expenses claims. The critical point in the system occurs when paper invoices need to be fed into the electronic system, by means of scanning. Paper receipts, important for the history of payment, are also scanned. As described, the point of scanning is a point at which errors are likely to occur.

References

Alvesson, M. (2009) "At-home Ethnography: Struggling with Closeness and Closure," in S. Ybema, D. Yanow, H. Wels, and F. Kamsteeg (eds.) *Organizational Ethnography*, London: Sage Publications.

Ancona, D. and Chong, C.-L. (1996) "Entrainment: Pace, Cycle, and Rhythm in Organizational Behavior," *Research in Organizational Behavior*, 18: 251–284.

Ancona, D., Goodman, P., Lawrence, B. and Tushman, M. (2001) "Time: A New Research Lens," *Academy of Management Review*, 26(4): 645–663.

Bechky, B. (2003) "Sharing Meaning Across Occupational Communities: The Transformation of Understanding on a Production Floor," *Organization Science*, 14 (3): 312–330.

Brown, J. S. and Duguid, P. (1991) "Organizational Learning and Communities-of-Practice: Toward a Unified View of Working, Learning, and Innovation," *Organization Science*, 2 (1): 40–57.

Bruner, J. (1986) *Actual Minds, Possible Worlds*. London: Harvard University Press.

Carlile, P. (2002) "A Pragmatic View of Knowledge and Boundaries: Boundary Objects in New Product Development," *Organization Science*, 13 (4): 442–455.

Conley, H. (2008) "The Nightmare of Temporary Work: A Comment on Fevre," *Work, Employment and Society*, 22 (4): 731–736.

Cooper, L. (1997) "Listening Competency in the Workplace: A Model for Training," *Business Communication Quarterly*, 60 (4): 75–84.

DeFillippi, R. and Arthur, M. (1998) "Paradox in Project-Based Enterprise: The Case of Film Making," *California Management Review*, 40 (2): 125–139.

Dholakia, U., Blazevic, V., Wiertz, C. and Algesheimer, R. (2009) "Communal Service Delivery: How Customers Benefit From Participation in Firm-Hosted Virtual P3 Communities," *Journal of Service Research*, 12 (2): 208–226.

Duguid, P. (1996) "Material Matters: The Past and Futurology of the Book," in G. Nunberg (ed.) *The Future of the Book*, Berkeley, CA: University of California Press. Available online at: http://people.ischool.berkeley.edu/~duguid/SLOFI/Material_Matters.htm. Accessed 6.7.2012.

Edmondson, A. (2004) "Learning from Mistakes is Easier Said than Done: Group and Organizational Influences on the Detection and Correction of Human Error," *Journal of Applied Behavioral Science*, 40 (1): 66–90.

Ewenstein, B. and Whyte, J. (2009) "Knowledge Practices in Design: The Role of Visual Representations as Epistemic Objects," *Organization Studies*, 30 (1): 7–30.

Forsey, M. (2010) 'Ethnography as Participant Listening" *Ethnography*, 11 (4): 558–572.

Gherardi, S. and Nicolini, D. (2002) "Learning the Trade: A Culture of Safety in Practice," *Organization*, 9 (2): 191–223.

Håland, E. and Tjora, A. (2006) "Between Asset and Process: Developing Competence by Implementing a Learning Management System," *Human Relations*, 59 (7): 993–1016.

Harper, D. (2003) "Framing Photographic Ethnography: A Case Study," *Ethnography*, 4 (2): 241–266.

Henderson, K. (1991) "Flexible Sketches and Inflexible Data Bases: Visual Communication, Conscription Devices, and Boundary Objects in Design Engineering," *Science, Technology & Human Values*, 16 (4): 448–473.

Hutchinson, V. and Quintas, P. (2008) "Do SMEs do Knowledge Management?: Or Simply Manage What They Know?," *International Small Business Journal*, 26 (2): 131–154.

Jacobs, C. and Coghlan, D. (2005) "Sound from Silence: On Listening in Organizational Learning," *Human Relations*, 58 (1): 115–138.

Jones, S. (2003) "Challenges of Developing a Sanitation Infrastructure Gis for the Tohono O'Odham Nation" *Public Works Management and Policy*, 8 (2): 121–131.

Kuhn, T. (2006) "A 'Demented Work Ethic' and a 'Lifestyle Firm': Discourse, Identity, and Workplace Time Commitments," *Organization Studies*, 27 (9): 1339–1358.

Lakoff, G. and Johnson, M. (1980) *Metaphors We Live By*, London: The University of Chicago Press.

Lamond, D. (1997) "Humanizing the Human Resource Planning Process: HRD at Universitas Terbuka," *Asia Pacific Journal of Human Resources*, 35 (1): 90–100.

Lave, J. and Wenger, E. (1991) *Situated Learning: Legitimate Peripheral Participation*, New York, NY: Cambridge University Press.

Lefebvre, H. (1991 [1947]) *Critique of Everyday Life*, Vol. 1, trans. by J. Moore, London: Verso.

Lefebvre, H. (2004 [1992]) *Rhythm Analysis, Space, Time and Everyday Life*, trans. by S. Elden and G. Moore, London: Continuum.

Levy, D. (2001) *Scrolling Forward: Making Sense of Documents in the Digital Age*, New York: Arcade Publishing Inc.

Macpherson, A. and Clark, B. (2009) "Islands of Practice: Conflict and a Lack of 'Community' in Situated Learning," *Management Learning*, 40 (5): 551–568.

Magaudda P. (2011) "When Materiality "Bites Back". Digital Music Consumption Practices in the Age of Dematerialization," *Journal of Consumer Culture*, 11 (1): 15–36.

Mead, G. H. (1934) *Mind, Self, and Society*, Chicago: University of Chicago Press.

Michlewski, K. (2008) "Uncovering Design Attitude: Inside the Culture of Designers," *Organization Studies*, 29 (3): 373–392.

Negroponte, N. (1995) *Being Digital*. New York: Alfred A. Knopf Inc.

Nicolini, D. (2009) "Zooming In and Zooming Out: A Package of Method and Theory to Study Work Practices," in Ybema, S., Yanow, D., Wels, H. and Kamsteeg, F. (eds.) *Organizational Ethnography*, London: Sage Publications.

Orlikowski, W. (2006) "Material Knowing: The Scaffolding of Human Knowledgeability," *European Journal of Information Systems*, 15 (5): 460–466.

Pantzar, M. and Shove, E. (2010) "Understanding Innovation in Practice: A Discussion of the Production and Re-production of Nordic Walking," *Technology Analysis and Strategic Management*, 22 (4): 447–461.

Patriotta, G. (2003) "Sensemaking on the Shop Floor. Narratives of Knowledge in Organizations," *Journal of Management Studies*, 40 (2): 349–375.

Pinch, T., Collins, H. and Carbone, L. (1997) "Cutting up Skills: Estimating Difficulty as an Element of Surgical and Other Abilities," in S. Barley and J. Orr (eds.) *Between Craft and Science. Technical Work in U.S. Settings*, Ithaca, NY: Cornell University Press.

Polkinghorne, D. (1988), *Narrative Knowing and the Human Sciences*, Albany: State University of New York Press.

Porter, T. and Córdoba, J. (2009) "Three Views of Systems Theories and their Implications for Sustainability Education" *Journal of Management Education*, 33 (3): 323–347.

Prentice, R. (2007) "Drilling Surgeons: The Social Lessons Embodied Surgical Learning," *Science, Technology and Human Values*, 32 (5): 534–553.

Rämö, H. (2002) "Doing Things Right and Doing the Right Things: Time and Timing in Projects,"*International Journal of Project Management*, 20 (7): 569–574.

Schon, D. and Wiggins, G. (1992) "Kinds of Seeing and Their Function in Designing," *Design Studies*, 13 (2): 135–156.

Sellen, A. and Harper R. (2003) *The Myth of the Paperless Office*, Cambridge MA: MIT Press.

Shove. E. and Pantzar, M. (2006) "Fossilization," *Ethnologia Europaea, Journal of European Ethnology*, 35 (1–2): 59–63.

Star, S. L. and Griesemer, J. (1989) "Institutional Ecology, 'Translations' and Boundary Objects: Amateurs and Professionals in Berkeley's Museum of Vertebrate Zoology, 1907–39," *Social Studies of Science*, 19 (3): 387–420.

Staudenmayer, N., Tyre, M. and Perlow, L. (2002) "Time to Change: Temporal Shifts as Enablers of Organizational Change," *Organization Science*, 13 (5): 583–597.

Toffler, A. (1970) *Future Shock*, New York: Bantam Books.

Trentman, F. (2009) "Materiality in the Future of History: Things, Practices and Politics," *Journal of British Studies*, 48: 283–307.

Volkema, R. J. and Niederman, F. (1996) "Planning and Managing Organizational Meetings: An Empirical Analysis of Written and Oral Communications," *The Journal of Business Communication*, 33 (3): 275–296.

Williams, R. (1974) *Television, Technolgy and Cultural Form*, New York: Shocken Books.

Yates, J. and Orlikowski. W. (2007) "The PowerPoint Presentation and its Corollaries: How Genres Shape Communicative Action in Organizations," in M. Zachry and C. Thralls (eds.) *Communicative Practices in Workplaces and the Professions: Cultural Perspectives on the Regulation of Discourse and Organizations*, New York: Baywood Publishing Company.

Ybema, S., Yanow, D., Wels, H. and Kamsteeg, F. (2009) "Studying Everyday Organizational Life," in S. Ybema, D. Yanow, H. Wels, and F. Kamsteeg (eds.) *Organizational Ethnography*, London: Sage Publications.

Zeiss, R. and Groenewegen, P. (2009) "Engaging Boundary Objects in OMS and STS? Exploring the Subtleties of Layered Engagement," *Organization*, 16 (1): 81–100.

Part III
Sharing and circulation

6 Practices, movement and circulation

Implications for sustainability

Allison Hui

Somewhere in Southern India, a man sits in a rickshaw. His long white beard proclaims his age at the same time that his accent and immaculate linen clothes belie his displacement. He's not from around here. The name of his website, www.haveyoga-willtravel.com, suggests that being away from his home in Australia is quite normal. Perhaps such trips have been occurring for years. On this day, the day he is being recorded for a video documentary, David Roche has travelled to India for a special occasion. It is the ninetieth birthday of his guru, Shri K. Pattabhi Jois, the founder of Ashtanga yoga. Sitting in the rickshaw, after having likely flown many miles, he tells an absent audience: 'Everybody that's in the Ashtanga world, pretty much, is here' (Wilkins 2006). He is only one of hundreds of yoga students from locations near and far to have made the trip.

This mass migration of yoga students contrasts sharply with the low levels of travel and resource consumption that are needed for everyday yoga practice. Unlike skiing or surfing, people don't need to leave home to participate – they can get on a yoga mat anywhere and start moving through the postures. Why then does someone like David Roche need to undertake long-distance travel? By examining the intersection of practices and travel more closely, this chapter suggests answers and, alongside them, new approaches for addressing environmental sustainability.

In the UK and elsewhere, reducing travel-related emissions is vital in moving towards a more sustainable, lower-carbon society. In England, between 2008 and 2011, the government invested over £43 million in the Cycling Cities and Towns programme, which aimed to get more people cycling (Redfern 2011). Charitable organizations such as the Ramblers and World Wildlife Fund, Canada have similarly headed campaigns that encourage people to engage in short-term experiments with sustainable modes of transport, in order to spark long-term changes ('Get walking keep walking' 2009; World Wildlife Fund Canada 'living planet community: pin it for the planet' nd). While such campaigns have had limited success (e.g. Sloman *et al.* 2010), they are typically grounded in a particular view of travel. This is one in which travel is taken to be an outcome of individual choice. Changing people's choices is seen as the best means of reducing travel-related emissions.

What such approaches fail to recognize is that patterns of contemporary travel are influenced not only by people, but crucially by the practices in which they participate. In other words, destinations and journeys are not arbitrary, nor are they simply expressions of choice: as argued here, they are better understood as outcomes of the specific ordering and organization of practices. Developing understandings of the links between practices and the movement of people is therefore important for ongoing attempts to manage patterns of mobility and reduce the carbon emissions associated with them.

This approach to explaining travel has a number of distinctive features. First, it foregrounds practices over individuals and, by doing so, acknowledges how travel patterns are shared and collectively enacted. It was not just David Roche who visited India for Jois' birthday – he was joined by hundreds of members of the global Ashtanga yoga community. Yoga is special in this way: similar trips to celebrate the birthday of a famous skier or surfer are uncommon. The shared travel within practices is therefore worth consideration. Second, such an approach recognizes that people's activities are intertwined with objects and knowledge. As elaborated in the next section, practices are not only about people. Finally, focusing upon practices as analytical units brings an additional spatio-temporal dimension into considerations of contemporary travel. Tracing where and when people travel has been a concern in many discussions of sustainability. Treating practices like Ashtanga yoga as units of study forces us to go further, asking how the practices themselves, and their circulation, have changed over space and time.

This chapter draws on two case studies, of Ashtanga yoga[1] and of leisure walking, to argue that the circulation of social practices mediates the movement of people. To put this point more concretely, it suggests that how David Roche ended up at a birthday party is connected to how Ashtanga yoga spread from India to the rest of the world.

Ashtanga yoga and leisure walking both have longstanding and global histories – a feature which facilitates consideration of their circulation to new spaces over time. Unlike surfing, which requires one of a limited number of aquatic settings, Ashtanga yoga and leisure walking can occur in different landscapes and locations. The possible spaces for performing are therefore more numerous than for other practices. Within this flexible and voluntary context, the enactment of distinct and collective travel patterns is especially relevant. In addition, the ambivalent contributions of these practices to travel-related carbon dioxide emissions – both have the potential to be either higher or lower impact – makes them particularly instructive cases.

Before getting into the detail of these cases, the next section introduces and defines the concepts of practices, movement and circulation. The following section uses these concepts to illustrate how the increasing circulation of Ashtanga yoga over time was related to specific patterns of movement. By comparing this case with leisure walking, the third section shows how different patterns of movement come to matter in diverse practices. Finally, the chapter ends by considering the implications of circulation for efforts to address the resource intensity of different practices.

Practices, movement and circulation

In everyday use, the term 'practice' is widely associated with what people do – indeed, it often communicates a general understanding of human activity. Theories of practice, however, are concerned not only with what *people* do, but also with what *practices* do. Bourdieu and Giddens, for instance, see practices as themselves generative – enacting through their repetition the structures of societies (Bourdieu 1977, 1990; Giddens 1979, 1984). This basic theoretical position, which has been taken up by subsequent scholars, leads to a methodological emphasis upon practices as the key units of study for social analysis.

A practice can be seen as 'a routinized type of behaviour' (Reckwitz 2002: 249). What distinguishes one practice from another is 'the distinctiveness of the *package* of doings and sayings plus organization that each is' (Schatzki 2002: 87). Practices can be understood in two ways – as performances that people enact in one time-space, and as entities that consist of the performances of many people in diverse time-spaces (Schatzki 1996: 89; Shove and Pantzar 2007: 154). Each time one person performs a practice, they reproduce its 'pattern' using 'a multitude of single and often unique actions' (Reckwitz 2002: 250). The total set of these reproductions makes up the practice-as-entity. Both Ashtanga yoga and leisure walking are entities in this sense, because they consist of recognizable collections of performances in multiple time-spaces.[2]

In order to perform, people must draw upon shared elements, which differ between practices. Within a practice, the vocabulary of elements includes:

> forms of bodily activities, forms of mental activities, 'things' and their use, a background knowledge in the form of understanding, know-how, states of emotion and motivational knowledge.
>
> (Reckwitz 2002: 249)

Ashtanga yoga, for instance, includes elements such as yoga mats, knowledge of yoga postures, embodied know-how for enacting these postures and motivations of calming the mind and stretching the body. Leisure walking, by contrast, often involves hiking boots, remembering to check the weather forecast, knowing how to read a compass or map and an appreciation of nature. Each time people perform these practices, they bring together some or all of the elements. Practices exist because of this coming together of people and distinct sets of elements.

In order to illustrate how these concepts relate to travel, additional distinctions are needed. While 'travel' can be used as a general term for displacement, the more specific concepts of 'circulation' and 'movement' are required to distinguish between the travel related to practices, elements and people.

The travel related to practice entities can be understood in terms of 'circulation'. Each practice entity is made up of many performances distributed in time and space, and the expansion and contraction of performances indicates the circulation of the practice entity. This circulation has two aspects – for one, it involves changes in the number of performances. When a practice is performed more frequently or by more people, its circulation increases, and when it is performed less frequently or by fewer people, it decreases. Circulation simultaneously involves changes in the spatial distribution of performances. The marked increase in yoga's circulation over the last century involved its spread beyond India (see Strauss 2005), while the falling circulation of games like marbles and jacks has led performances to occur in fewer spaces. In this way, speaking of the circulation of practice entities highlights the expansion or contraction, in number and distribution, of performances over time.

While circulation refers to the changing spatiality of the practice entity, the concept of movement can be used to isolate single trips of people and objects. Movement refers to the displacement of something in objective space (Cresswell 2006: 3). One yoga practitioner, such as David Roche, moves through certain spaces over time. His yoga mat has its own movements, which are both related and unrelated to his. Though the movements of different entities can be compared or amassed, movement is a characteristic of single units – in particular, of people and materials. Circulation, by contrast, cannot be disaggregated from its collective entity – a practice.

How then are the circulation of practices and the movement of people and elements interlinked? In order to perform a practice, and contribute to its circulation, people must amass the appropriate elements, and this often requires movement to buy things or learn skills. Elements themselves also move – yoga mats are made in factories and shipped to stores before being carried to classes. In order to explain how people and elements come to be in various places though, the relationship between circulation and movement needs to be considered further.

When discussing the time-space of human activity, Schatzki suggests that the past organization of practices 'circumscribes contemporary activities' because it provides a normative environment that people generally uphold (2010: 214, 211). Think, for instance, of places where there are rules against smoking in restaurants. The fact that previous patrons complied with this rule constrains current patrons, by making censure, or requests to leave, both likely and acceptable consequences of breaking the rule. Past performances therefore serve as a model that establishes precedents and delineates the implications and meanings of present performances.

A similar point can be made about the relationship between the circulation of practices and the movement of people and elements. The past circulation of a practice mediates contemporary movements of people and elements. The spatial distribution of performances in the past affects accessibility to elements in the present. Think, for instance, of collecting the elements of skiing. If you wanted to buy ski boots or find a teacher, visiting a large city in the tropics would be less successful than visiting a ski resort. While online commerce has shifted geographies of retail, elements of practice are more likely to be found near the people or spaces linked to past performances. The movement of elements can in this way become patterned and normalized, indicating the likely geographies of continued availability.

The following sections expand upon this point, demonstrating how specific types of movement become normalized and reproduced in Ashtanga yoga and leisure walking. They argue that movements of people and elements (1) are influenced by the context-specific circulation of practices and (2) reinforce practice-specific meanings and goals.

The discussion focuses on the expanding circulation of these practice entities from spatially limited beginnings to popular global pursuits. This focus obviously leaves out other important issues such as the contraction or fluctuation of practice entities at particular moments. Looking at expansion, however, provides an opportunity to explore the development of practice-specific patterns of movement, as well as links between circulation and movement.

Ashtanga yoga: Mysore and beyond

Today, Ashtanga yoga is a global practice. The website Ashtanga.com, for instance, lists upcoming workshops in over forty countries around the world, from Chile to Bhutan, and from Russia to Japan (Ashtanga yoga workshops 2011). Forty years ago, however, Ashtanga yoga was only performed in India. Though this expansion could be narrated in many ways, the stories of circulation told by practitioners foreground the importance of people's movement (Donahaye and Stern 2010; Jois 2002; Ryan 2009; Williams 2009a). They go something like this:

Even in the 1920s, yogis travelled. Yogi Krishnamacharya was moving through India in 1927, demonstrating and lecturing on yoga, when he encountered Shri K. Pattabhi Jois. Jois, who was only twelve at the time, was so inspired that he immediately asked to study with the older man. Krishnamacharya accepted him as a student and, in 1930, Jois became a travelling yogi, moving to Mysore to continue his study. In 1948, he made Mysore a more permanent home for Ashtanga yoga by establishing the Ashtanga Yoga Research Institute.

Several decades later, it was Jois' son Manju who was the travelling yogi, visiting Pondicherry in 1972 to give a demonstration of his father's Ashtanga yoga system. A young American named David Williams was in the audience that day, having come to India in search of the greatest yogi. Captivated by what he saw, he became determined to study with Jois. His search took him to Mysore for four months in 1973, after which he returned to California and began teaching the first international classes in Ashtanga yoga.

Conventional accounts go on to describe Jois' first trip to America in 1975. Jois visited with his son Manju, and after teaching together for several months Manju decided not to return to India with his father. Instead, he continued teaching in California, which allowed Williams to move to Hawaii and extend the circulation of Ashtanga even further. These three bases of Mysore, California and Hawaii subsequently became destinations for a new generation of yogis, including David Swenson and Danny Paradise, who have produced instructional videos and taught hundreds of students, among them celebrities such as Sting and Madonna (Naddermier 2010; White 1995; Williams 2009b).

These few details provide a basic narrative of how Ashtanga's circulation expanded – how more people came to perform the practice in more spaces, leading eventually to the global dispersion found today. To understand how Ashtanga's circulation mediates the movement of both people and elements, however, one needs to look further into the past. One needs to consider how yoga, and its requisite elements, circulated even before Ashtanga developed.

For thousands of years, yoga has been taught through guru-disciple relationships. In the early Vedic period (c.1200–900 BCE), young boys would spend years memorizing knowledge transmitted orally by a teacher (Whicher 1998: 35–36). Students would then grow up to be gurus, who would teach others:

> So it has been proceeding for an unknown number of centuries in India, for at least three millennia in the longest unbroken apostolic succession anywhere in the world. The *Upanishads* are nearly that old, and they contain lists of gurus who were taught by gurus, back to the beginnings of Hindu time.
>
> (McArthur 1986: 109)

This tradition of teaching within India was reproduced so consistently that it arguably became 'a self-evident fact that no Hindu would dream of questioning' (Varenne 1976: 93). To learn yoga was to be a disciple. The weight of this tradition informs Jois' repeated assertion of the need to study under a guru: 'Yoga should never be learned from reading books or looking at pictures. It should only be learned under the guidance of a Guru who knows the yogic science and is experienced in its practice' (Jois 2002: 28, see also 20, 59, 63, 65, 77). This system of apprenticeship and discipleship, however, comes with specific consequences for movement. Discipleship depends on co-presence; to maintain guru-student patterns of learning, people must travel.

In the case of Ashtanga, reproducing this tradition of discipleship was initially unavoidable. While today there are books and DVDs through which one can access the elements of Ashtanga yoga, it was not until 1999 that Jois' only book was translated into English (Jois 2002). Therefore, though English print media helped significantly in the circulation of other yoga teachings (Strauss 2005: 40–43), Ashtanga yoga remained dependent upon traditional guru-disciple relationships. The activities, skills and knowledge of Ashtanga yoga could only move beyond India with people, and so where teachers went and who they taught became vitally important for the expansion of the practice entity.

Though some contemporary participants do not follow guru traditions, the stories that narrate the circulation of Ashtanga yoga continue to reinforce an idea of how Ashtanga should expand, thus normalizing the movement of people. Further, to maintain these guru traditions, the independent movement of resources (such as books) is delegitimized: 'Can a book replace the guru, the master? Not completely, of course' (Svatmarama and Rieker 1992: 14). The guru tradition promotes the transmission of knowledge, while also limiting it through the stipulation that elements, especially of skill, should move in tandem with people.

Therefore, people's movement continues. Among senior students and teachers, travel remains an important marker of devotion to a guru and to the practice (Smith 2004: 8). Students still flock to Mysore in order to further their skills because others have done so before them: 'Mysore to an Ashtanga practitioner is like Hawaii is to a surfer. It is the source. It is where all the most serious Ashtanga yoga practitioners have learnt' (Brundell nd). This movement is actively encouraged by the Ashtanga Yoga Research Institute, which is now called the Shri K. Pattabhi Jois Ashtanga Yoga Institute (KPJAYI). As the only organization giving official teacher certification, it requires teachers to 'study regularly at the KPJAYI' and, during 2008 and 2009, even specified that top level instructors were to make eight annual trips (Shri K Pattabhi Jois Ashtanga Yoga Institute: Teachers Information 2009). In order to be a true enthusiast of Ashtanga, one must seek out opportunities for co-presence with gurus. Thus, though Ashtanga can be performed in a variety of spaces, its circulation reinforces the aphorism on David Roche's website – have yoga, will travel.

In order to demonstrate that this mediation of personal movements by practices is not particular to Ashtanga yoga, the next section looks at the case of leisure walking. As it shows, where and how people and elements move is similarly structured by the circulation of the practice entity, but in ways that encourage very different patterns of travel.

Leisure walking on the move

At first glance, it might appear that the circulation of Ashtanga yoga and leisure walking are quite similar. Both developed from older practices – hatha yoga and walking as transportation – and grew from an initially small group of practitioners to become widespread global activities. In addition, both have been connected to lineages. While Jois is the guru of the Ashtanga tradition, Wordsworth was among those who 'founded the whole lineage of those who walk for its own sake and for the pleasure of being in the landscape' (Solnit 2001: 82). Yet looking more closely at how these practices are passed on, it becomes clear that their circulation was supported by different patterns of movement.

Perhaps Wordsworth was lonely on his birthday. After all, many of those in the lineage he started never met or visited him. He was never a privileged interpreter, directly passing on knowledge about how to walk for pleasure. This is because while walking in the country 'is beset by conventions about what constitutes "appropriate" bodily conduct, experience and expression' (Edensor 2000: 82–83), there is little ethical prohibition against learning these from a variety of sources by observation, trial and error. One need not seek out a guru of walking when it is possible to figure it out independently. The lineage Wordsworth headed was therefore never responsible for teaching the embodied activities of walking. His lineage was imagined, not co-present. It did not come with the obligatory personal travel of discipleship.

Instead, his poems and essays gave the everyday activity of walking new meanings and motivations. In Wordsworth's hands, walking and wandering are re-defined, invested with notions of return and positive energy that supersede previous ideas about wandering as a dangerous threat to community (Wallace 1993: 120, 122). Even though people did not travel to study with him, books communicating his experiences of walking moved widely, providing easy access to the understandings, emotions and motivations that supported his performances. Books were, in a sense, able to become a part of the lineage, transmitting elements as they travelled from author to reader.

It was not only Wordsworth's books that travelled, since he was one of a number of walking writers. Solnit places Rousseau at the beginning of the series of poets and essayists who frame 'walking as a conscious cultural act rather than a means to an end' (2001: 14). Though these authors used differing justifications for the value of walking, together they transformed ideas by becoming its advocates (Solnit 2001: 82). Wordsworth was central among these writers, because walking was such an obvious theme in his work, as well as a key part of his daily life and method of composition. Multiple authors in multiple countries, however, provided a repertoire of slightly varying understandings of the practice, its goals and its motivations.

Unlike Ashtanga yoga, the elements of leisure walking originated in many places, and routinely travelled separately, not tied to any one practitioner. As a result, the expansion of leisure walking as a practice entity occurred in a more geographically distributed manner, and without strictly patterned movements of people.

While people weren't obliged to visit certain spaces in order to learn, their movement was mediated by writing walkers, who highlighted appropriate spaces for leisure walking. Wordsworth, for instance, lauded the places in which he walked, contributing in particular to the creation of the Lake District as a destination for visitors. As Urry notes, 'there is nothing obvious or inevitable about why huge numbers of people would voluntarily choose to visit [the Lake District], a place that up to the eighteenth century was seen as the very embodiment of inhospitality' (1995: 193). The writing of Wordsworth, however, championed the virtues of walking in such places, and encouraged visits by depicting the Lake District as an appropriate space to perform the desires and activities of walking (Urry 1995: 200). Similarly, Snowdon, Cader Idris and Plynlimmon were all popular ascents in the late eighteenth and early nineteenth centuries, largely because of Coleridge and Wordsworth's writings (Marples 1959: 109). By linking leisure walking to particular spaces, these poets shaped people's movements, investing certain destinations and trips with significant pleasure and esteem.

Though writers discussed appropriate spaces for walking, the diversity of these spaces reinforced the point that no one was essential. As Wallace notes, Wordsworth emphasized that it was important 'not to gain knowledge of any particular place or places, but to be able to see and examine the moving passage between places, the process of change itself' (Wallace 1993: 73; Wordsworth 1977: appendix 2). Jarvis similarly suggests that while Wordsworth has become associated with the Lake District and Grasmere, he was 'remorselessly addicted to travel' (2001: 321). In this way, the idea that leisure walking should be performed in many spaces was also transmitted through his writings. As a result, the circulation of leisure walking did not depend upon specific spaces in the same way that Ashtanga yoga did. While Mysore is considered the 'source' of Ashtanga yoga (Brundell nd), leisure walking involves many notable spaces: a 'greatest-hits tour' (Odell 2007). This understanding of the importance of visiting multiple spaces continues to be re-produced and re-enacted in people's diverse movements today.

The circulation of leisure walking, however, is not about writers alone. Though there were many spaces where leisure walking could occur, including the wild and natural spaces that these writers legitimized, people still had to reach them. People's movements therefore need to be considered more broadly in relation to the positioning of walking.

Before the transportation revolution started in the late eighteenth century, people travelled primarily by foot. However, during the eighteenth century, roads were poorly constructed and frequented by highwaymen who would attack travellers. Those with money acquired horses or carriages to avoid walking because, in addition to being dangerous, it was heavily stigmatized: 'To walkers and non-walkers alike, walking as travel meant poverty, alienation from society whether for legal or extra-legal reasons, possible moral turpitude, and probably danger to the individuals and communities touched by the act' (Wallace 1993: 33). The development of railways was therefore remarkable because it provided a new, affordable means of transportation and facilitated the reframing of walking as a choice.

This new transportation infrastructure was crucial to the circulation of leisure walking because it allowed meanings of walking to change. Walking was no longer a signal of low socio-economic status or degeneracy: 'since the common person need not travel by walking, so walking travellers need not necessarily be poor' (Wallace 1993: 62). In addition to opening up understandings of walking, rail travel was much safer and faster than other modes of transportation, and encouraged walking tours by increasing accessibility to the countryside. Freed from the inconveniences of walking as travel, people could take the railway to places in which they wanted to walk, without having to cover the entire distance on foot (Wallace 1993: 66). In this way, more and more spaces away from local communities became possible places for walking. By changing the possibilities for how people travel, railways introduced shifts into the movement of both people and ideas about transportation. They thereby contributed to the expansion of leisure walking from 'a leisure activity of the aristocracy to a popular pursuit of the upper middle classes' (Amato 2004: 101). The movement of the objects and technologies of rail travel worked alongside that of writings to facilitate the circulation of leisure walking; both via an increased number of performances, and a wider spatial (and social) distribution.

In summary, these elements – the understandings, motivations and goals that distinguish leisure walking from everyday walking – were developed in many spaces by many people, and were shaped by broader changes in transportation technologies. Their early patterns of movement continue to resonate and be reproduced today. Books and writings remain dominant resources, which suggest the value of travelling to many spaces rather than a select few. The separation that railway travel enabled between everyday walking and leisure walking also remains relevant, with leisure walking being primarily associated with non-urban spaces. As this case shows, shifts and changes in one mode of transportation can have implications for the movement and circulation of other elements and practices.

These cases demonstrate that the circulation of practice entities both shapes and is shaped by the movement of elements and people. Past destinations, justifications and motivations circumscribe contemporary movements. Past movements that have helped to expand the circulation of a practice become valued modes of transmission, and models for the future. Some types of movement come to matter more than others because they have been proved to be effective in the past. In Ashtanga yoga, the movement of people is important because it provides an effective way of transmitting knowledge and skills, and of reproducing understandings of lineage and discipleship. For leisure walking, the decentralized movement of ideas and understandings through books became essential to the practice's circulation. These types of movements form precedents that have implications for contemporary performances. Moving long distances to meet with gurus is legitimated and encouraged in Ashtanga yoga, while the independent and widespread movement of books and people is normalized in leisure walking.

Though these movements can always change, and must be re-enacted to persist, the point here is that they come to have meaning, and are patterned by the practices of which they are a part. Imagining the inversion of current patterns of movement – suggesting leisure walking enthusiasts must show devotion by visiting the Lake District in England, or that Ashtanga yogis figure out postures independently – further emphasises how practices mediate movement. For individual practitioners, such mediations mean that certain patterns of movement can be unintended consequences of pursuing specific practices. Beginning yogis may not set out to travel, but find that as they progress, studying with particular gurus becomes important to increase their skills and support core ideals of the practice. Movements and their meanings can in this way emerge over the course of people's participation, just as particular patterns and traditions of movement emerge over the course of a practice's circulation.

Conclusion: practice-mediated travel and intervention

This chapter has shown how instances and patterns of travel, rather than simply being the outcome of individual choice, are mediated by social practices. Across the historical trajectory of a practice's development, particular spaces and movements of people and objects become normalized. The circulation of practices shapes the continued patterns of movement that support them. To conclude, I take these ideas and reflect on their potential to reframe sustainable travel policy and open up new avenues for intervention.

A theme, by now familiar, is that patterns of movement are the result of participation in practices. In this sense, travel is somewhat unintentional. It is, therefore, no wonder that efforts based solely on informing people about carbon miles and better choices have met with limited success (Southerton *et al.* 2011: 17). Though individuals might be concerned about their carbon footprints, information about cutting carbon miles will hold little sway if the performance of a practice requires them. From this point of view, small targeted projects that attend to specific practice-mediated patterns of travel could prove more effective. Sustainable travel policies, rather than attempting to speak to wide sectors of the population, could be tailored to particular, travel-intensive, practices.

Specifying which practices these are is beyond the scope of this chapter, but briefly speculating on the cases presented here, the list would likely be broad-ranging, and have unexpected contenders, such as yoga, in its midst. Such an approach to intervention would need to weigh up the potential trade-offs of reducing the 'moving things' of practices. That is to say, eliminating some movements may initiate others; for instance, reducing yogis' travel to India might increase the movement of books and DVDs. Evaluating the overall impacts of practice-related travel would need to take account of such dynamics.

This chapter also highlights that a practice's circulation is always changing, and as such, associated patterns of travel are also always 'on the move'. For example, before the 1970s, Ashtanga yoga had low levels of travel-related resource use because students didn't have to travel far to see their guru. As the practice circulated internationally, long-distance trips became important for maintaining student-teacher relationships. Global circulation does not always prompt such increases in resource use; although leisure walking is a worldwide pursuit, international travel is less important. However, in both cases, the travel implications of being a practitioner shift along with the growth and expansion of the practice.

These observations illustrate that the frequencies and distances of travel associated with a practice are dynamic. They have been different in the past, and they will be different in the future. The role of intervention, then, is to shift the practice to a more sustainable trajectory as part of this ongoing transformation. Acknowledging that practice-mediated travel is always shifting, means that there is scope for identifying and targeting undesirable trajectories early, before they take hold. The variety of legitimate spaces associated with leisure walking might offer some clues here. Wordsworth's emphasis upon walking in many spaces, rather than just a select few, helped to justify a wide range of sites for walking, making local participation possible. Emphasizing variety and the appropriateness of many sites for performance might be one potential strategy. Such efforts, however, would need to be attentive to the potential link between variety and a travel-intensive push towards collecting and visiting all possible sites.

Finally, as this latter point suggests, intervening in practice-mediated travel cannot be accomplished with a 'business as usual' attitude. Any intervention or change will result in winners and losers. As travel becomes integrated into performance, it comes to have meaning and value to practitioners as well as contributing to tourism revenue (Strauss 2005 shows the importance of yoga for India's tourism). Reconciling the goals of a lower-carbon society with contemporary practices will be a contested process, and may depend upon simultaneous changes to practices and how societal wealth and benefits are evaluated (see Sayer, Chapter 11, for a more detailed discussion of these issues).

In this sense, the contribution of this chapter is simply a first step, illustrating just how embedded movement has become in social life and identifying a potential reframing of intervention that takes us beyond behaviour change and into the complexities of social practice.

Acknowledgements

This chapter draws upon research supported by a PhD scholarship from the Commonwealth Scholarship Commission of the UK. Many thanks to Elizabeth Shove, Nicola Spurling, John Urry, Dana Bentia, Misela Mavric, James Tomasson, Niklas Woermann and the other contributors for their extremely helpful comments on earlier drafts.

Notes

1 This is a type of hatha yoga, in which a series of postures are sequenced with specific gazing points and breaths: see for example the diagrams at www.yoga-manchester.co.uk/mats-yoga-charts.pdf.
2 Though defining practices is always an analytic task, the existence of professional or voluntary organizations in each of these cases, along with the ability of practitioners to name and identify with these pursuits, provides a strong justification for their validity as distinct practices.

References

Amato, J. A. (2004) *On foot: a history of walking*, New York: New York University Press.

Ashtanga yoga workshops (2011) Available online at: www.ashtanga.com/workshops.lasso (accessed 11 April 2011).

Bourdieu, P. (1977) *Outline of a theory of practice*, trans. R. Nice, Cambridge: Cambridge University Press.

Bourdieu, P. (1990) *The logic of practice*, trans. R. Nice, Stanford: Stanford University Press.

Brundell, B. (nd) 'Mysore', available online at: www.planetashtangayoga.com/Mysore.html (accessed 25 April 2012).

Cresswell, T. (2006) *On the move: mobility in the modern Western world*, London: Routledge.

Donahaye, G. and Stern, E. (2010) *Guruji: a portrait of Sri K. Pattabhi Jois through the eyes of his students*, New York: North Point Press.

Edensor, T. (2000) 'Walking in the British countryside: reflexivity, embodied practices and ways to escape', *Body and Society*, 6(3–4): 81–106.

Gaddi, R. and Bellavitis, A. D. A. (2010) 'Design, knowledge sharing, creativity, great events: tools for contemporary urban development', *Revista de Design, Inovação e Gestão Estratégica*, 1(1): 96–112.

'Get walking keep walking' (2009) Available online at: www.getwalking.org/ (accessed 15 February 2012).

Giddens, A. (1979) *Central problems in social theory*, London: Macmillan Press.

Giddens, A. (1984) *The constitution of society*, Cambridge: Polity Press.

Jarvis, R. (2001) 'The wages of travel: Wordsworth and the memorial tour of 1820', *Studies in Romanticism*, 40(3): 321–343.

Jois, S. K. P. (2002) *Yoga mala*, trans. S. V. Kadam and D. H. L. Chandrashekar, New York: North Point Press.

Lerner, J. (2007) 'Cities, sociodiversity, and strategy', in M. Robinson, W. Novelli, C. Pearson and L. Norris (eds), *Global health and global aging* (pp. 275–280), San Francisco: Wiley.

McArthur, T. (1986) *Understanding yoga: a thematic companion to yoga and Indian philosophy*, Wellingborough, UK: The Aquarian Press.

Marples, M. (1959) *Shanks's pony: a study of walking*, London: J. M. Dent & Sons Ltd.

Naddermier, K. (2010) 'Conversations with Danny Paradise', available online at: http://yogaparis.wordpress.com/2010/05/07/conversations-with-danny-paradise/ (accessed 25 April 2012).

Odell, R. (2007) 'Life list: backpack Rocky Mountain National Park', *Backpacker*, 35(253): 30.

Reckwitz, A. (2002) 'Toward a theory of social practices: a development in culturalist theorizing', *European Journal of Social Theory*, 5(2): 243–263.

Redfern, R. (2011) 'Evaluation of the cycling cities and towns programme: interim report', London: Department for Transport.

Ryan, M. (2009) 'The boy David', *Yoga Magazine*, 76: 56–58.

Schatzki, T. R. (1996) *Social practices: a Wittgensteinian approach to human activity and the social*, New York: Cambridge University Press.

Schatzki, T. R. (2002) *The site of the social: a philosophical account of the constitution of social life and change*, University Park, PA: Pennsylvania State University Press.

Schatzki, T. R. (2010) *The timespace of human activity: on performance, society, and history as indeterminate teleological events*, Lanham, MD: Lexington Books.

Shove, E. and Pantzar, M. (2007) 'Recruitment and reproduction: the careers and carriers of digital photography and floorball', *Human Affairs*, 17: 154–167.

Shri K Pattabhi Jois Ashtanga Yoga Institute (nd) Teachers, available online at: www.kpjayi.org/teachers.html (accessed 6 November 2009).

Shri K. Pattabhi Jois Ashtanga Yoga Institute (2009) Teachers Information, available online at: http://kpjayi.org/the-institute/teachers (accessed 31 March 2011).

Sloman, L., Cairns, S., Newson, C., Anable, J., Pridmore, A. and Goodwin, P. (2010) 'The effects of smarter choice programmes in the Sustainable Travel Towns: summary report', London: Department for Transport.

Smith, B. R. (2004) *Adjusting the quotidian: ashtanga yoga as everyday practice*, paper presented at the Cultural Studies Association of Australia Conference, Perth, 9–11 December.

Solnit, R. (2001) *Wanderlust: a history of walking*, London: Verso.

Southerton, D., McMeekin, A. and Evans, D. (2011) *International review of behaviour change initiatives: climate change behaviours research programme*. Edinburgh: Scottish Government Social Research.

Strauss, S. (2005) *Positioning yoga: balancing acts across cultures*, Oxford: Berg.

Svatmarama, Y. S. and Rieker, H.-U. (1992) *Hatha yoga pradipika*, trans. E. Becherer, London: Aquarian Press.
Urry, J. (1995) *Consuming places*, New York: Routledge.
Varenne, J. (1976) *Yoga and the Hindu tradition*, trans. D. Coltman, Chicago: University of Chicago Press.
Wallace, A. D. (1993) *Walking, literature, and English culture: the origins and uses of peripatetic in the nineteenth century*, Oxford: Clarendon Press.
Whicher, I. (1998) *The integrity of the yoga darśana: a reconsideration of classical yoga*, New York: State University of New York Press.
White, G. (1995) 'Every breath you take: Sting on yoga', *Yoga Journal* 125: 63–69.
Wilkins, R. (2006) (Director) 'Guru' [Documentary Film].
Williams, D. (2009a) 'Lecture on Ashtanga yoga', Dancehouse Theatre, Manchester, 18 June.
Williams, D. (2009b) 'Lecture on the history of yoga', Dancehouse Theatre, Manchester, 20 June.
Wordsworth, W. (1977) *Guide to the lakes*, Oxford: Oxford University Press.
World Wildlife Fund Canada (nd) 'living planet community: pin it for the planet', available online at: https://community.wwf.ca/PinIt/index.cfm (accessed 15 February 2012).

7 Sharing conventions

Communities of practice and thermal comfort

Russell Hitchings

Introduction

There is much current interest in harnessing 'communities' to make more sustainable societies. We could understand this interest in strategic terms as the purposeful transfer of responsibility for reducing carbon emissions from governments to the public (Seyfang 2008). We could also make a more positive case about the potential for reducing energy demand at the same time as fostering social cohesion (Connors and McDonald 2011). Either way, such discussions generally deploy a concept of 'community' defined by geography. The communities referred to consist of people who live near one another, such as those in rural locations who have worked together to adopt renewable energy technologies (Walker *et al.* 2007), or those in cities who are being encouraged to develop similar schemes (Aiken 2012). This framing is attractive, because it chimes with ideas about 'transition', which position spatially bounded communities as effective crucibles of sustainable change (Seyfang and Smith 2007) and with romantic beliefs about those living together naturally enjoying a strong sense of connection (Franklin *et al.* 2011). Yet it may not always be those living nearby who influence practices, including those that matter for energy use. Many of us barely know our neighbours and feel much closer to other reference groups. We might therefore benefit from conceptualising these collectives in ways other than those defined by geographic proximity.

This chapter considers how aspects of a 'communities of practice' framework might help promote sustainable living. Within this approach, 'community' no longer revolves around proximity, but centres instead on participation in identified practices, understood as relatively widespread ways of doing things. It draws particular inspiration from Lave and Wenger (Lave and Wenger 1991; Wenger 1998) and their anthropological interest in the circumstances under which people learn to emulate and in how practices change through interaction between newcomers and old hands. In their account of social practice, Lave and Wenger are particularly alive to the situational processes through which individuals come to undertake tasks in similar ways. As such, the 'communities' they discern are both unstable and a matter of practical achievement. Furthermore, and perhaps most important for my discussion, establishing whether such communities exist or not is taken to be an empirical question. Developing this approach, I suggest that attending to the formation and prevention of practice communities could enrich understanding of mundane energy use. In short, my chapter explores how some groups develop as a 'community of practice', whilst others do not, and considers the implications of the processes involved for efforts to encourage less resource intensive modes of living.

I begin with a brief discussion of recent versions of social practice theory as applied to the sustainability challenge, focusing on the degree to which they frame practices as shared, and also on how the 'communities of practice' approach might add to this agenda. I then return to two previous research projects and consider whether communities of practice could be identified within the groups involved in these two studies. Though concerned with different contexts, both research projects shared an interest in seasonal climate change, the everyday achievement of thermal comfort, and the management of indoor temperature. The practices under scrutiny were, therefore, those involved in keeping human bodies sufficiently warm or cool. I start with a discussion of how professional office workers responded to the arrival of summer warmth at work and then consider how older people managed winter cold at home. Through this comparative exercise I explore the subtleties of how communities of practice are either consolidated or impeded from developing. The chapter ends by taking stock of the implications of these enquiries, focusing on the potential for sustainability 'brokering' and the relevance of attending to two different kinds of 'insulation'.

How do practice communities form?

Recent work applying social practice theory to sustainability agendas has criticised the policy focus on individualistic conceptions of action (Shove 2010). The argument here is that, because of their focus on individual choice, dominant approaches tinker around the edges of the problem. These critics contend that instead of focusing on individual behaviour we should concentrate on broader questions about how concepts of 'normal' (often resource intensive) ways of living evolve, and how these are powered by dynamics that have yet to be understood. This puts the spotlight on how, at given points in time, broader collectives establish and achieve everyday objectives, rather than on how individual traits drive behaviour. For example, Chappells *et al.* (2011), study how gardeners use water outside their homes and how watering practices have changed. Shove similarly examines changing ways of doing laundry and how tumble dryers have come to replace washing lines (Shove 2003). This emphasis on shared conventions – rather than personal attributes – leads to new ideas about how sustainable practices might be fostered.

Though useful in moving away from an assumption of significant personal agency, and from the view that individuals are free to choose more and less sustainable ways of life, this approach can sometimes suggest there are, at given points in time, relatively undifferentiated combinations of understandings, procedures, and engagements that define and govern appropriate conduct within any given practice (Evans 2011). By concentrating on the emergence or disappearance of practices, accounts like these are at risk of downplaying the significance of diversity and difference. Focusing on broader sweeps of change is valuable in that it underlines the importance of collective arrangements, but such an emphasis also means that less attention has so far been paid to variation in how practices are concurrently reproduced. In this chapter, I suggest that understanding variation in how practices are performed is relevant and necessary if we are to figure out how such variation might be encouraged or impeded to positive environmental ends.

In taking this approach, I make use of Lave and Wenger's discussions of how 'communities of practice' are forged and maintained (Lave and Wenger 1991; Wenger 1998). Their particular interest was in processes of 'social learning' and in how practices and skills are shared and reproduced in different settings. The fundamental point, for them, is that learning is not a discrete activity, associated with formalised spaces of teaching or isolated contemplation, but much more a matter of practical activity, as people learn from each other by mutually engaging in tasks (Lave 1993). Partly because their insights are grounded in a series of ethnographic studies, their characterisation of this process is particularly alive to the details of lived experience and the context dependent mechanisms through which communities of practice form. It is this idea that I want to take forward.

Wenger tells us that practices are defined by 'boundaries' (Wenger 1998: 121), such that those inside the practice community have a relatively clear sense of what constitutes a proper performance, and that this is not shared by those outside the boundary. Understood in this way, the shared nature of social practices becomes a matter of empirical investigation. Identifying processes of boundary maintenance and disruption are useful tools for this task. Rather than reifying generic processes of 'transmission' (Turner 1994), Wenger argues that we should focus on the situated creation and maintenance of practice community boundaries. Though the promotion of less energy intensive modes of existence was far from their concerns, I suggest that the concepts of practice community formation and boundary making, of the kind that Lave and Wenger extract from their studies of organisations and apprenticeship, could be used to identify new ways of encouraging sustainability.

Practices are, by definition, social, in the sense that they are shared and recognised by others. Indeed, this is one of the core reasons why the concept has proved attractive, since it helps guard against the temptation to drift back towards accounts of social life that place undue emphasis on individual autonomy (Reckwitz 2002). However, we should not assume practices are always performed in the same way. As stated, in this chapter I want to focus on variety and on how communities of practice take shape. In this sense, I am interested in the obstacles and experiences associated with practice 'recruitment' (Reckwitz 2002: 250), and with how processes of recruitment and defection influence the formation (and dissolution) of communities, whose members perform specific practices in relatively similar ways.

I now revisit two previous research projects, both focusing on methods of achieving thermal comfort and on whether strategies varied across the seasons. There are various ways in which humans keep themselves sufficiently warm or cool. These include bodily adaptation, clothing selection, and ambient temperature control. Some of these methods are much more energy intensive than others. In focusing on this topic I am interested in three related questions. First, under what conditions do people achieve thermal comfort in similar ways. Second, what are the most significant factors impeding or encouraging the emergence of similar strategies and practices. Third, does a discussion of how such practices are shared and how they vary reveal what is relevant for those interested in promoting sustainability?

Held inside a strengthening community of practice: forgetting about summer clothes in city offices

The first study that I discuss involved in-depth interviews with professional office workers in the financial heart of London. The research was, in part, designed to discover whether these people responded to summer in the same way: in short, did they form a community of practice in this respect? This is a distinctive group in being quite removed from the outdoor environment. Since these workers spend long periods of time inside climatically controlled offices, it was possible that they might be relatively oblivious to outdoor temperature change. This could show itself in a lack of seasonal fluctuation in clothing and an increasing reliance on air conditioning. Serial interviews with lawyers allowed me to discover whether this was so (see Hitchings 2011 for more detail). I now use this material to consider the extent to which members of this group shared methods of managing thermal comfort, to show how these varied with the seasons, and to identify what held this particular community of practice in place.

A common discussion topic related to the social 'arrival' of summer and whether respondents amended their habitual modes of dress, and weekday routines, in response to seasonal warmth outdoors. In many cases, they did not. When asked about the preceding summer, respondents were often forced to think on their feet. It was not unusual for interviewees to have little physical contact with the outdoor climate. This was generally seen to be both normal and acceptable. When I asked whether a respondent was experiencing a 'good summer', proxy indicators were often enlisted in formulating a response to a question that few had really considered before. For example, one respondent had been working hard towards the goal of securing a partnership at her firm. Techniques for coping with the weather were consequently low on her list of concerns. In passing, she noticed an unusually big dry-cleaning bill as a result of some recent rains, which had muddied the path that began her commute; but that was all. Other preoccupations pushed seasonally specific practices off the agenda. This was true of other respondents too, and, for most, this was not something that was experienced as being particularly problematic.

Sharing conventions 107

In thinking about how this community of practice formed, the tendency to disregard summer appeared to be amplified by various on-site services fostering indoor sociability. For example, the company canteen offered a meeting and eating space more convenient than other less proximate alternatives. Meeting colleagues or friends during breaks was commonly done in indoor environments, even during the summer. The presumption was that others might find it awkward (or unpleasant) to go outdoors, and that it was safer to arrange to meet in climatically controlled indoor spaces. In other words, it was better to suggest an appropriately cool location than run the risk of making peers uncomfortable. Respondents made exceptions when meeting with close friends, and when they were sure that their companion would be willing to cope, but even then, going out was often something of an event. Convention and a concern to conform to (presumed) expectations, resulted in an identifiable pattern of seasonally unchanging practice. The continued reproduction of such 'non-seasonal' arrangements meant that there was no need to dress differently in the summer, but, as we are about to see, some liked the idea of doing so.

On closer inspection, there was some variation in the clothing worn in these city centre offices. When we spoke of summer clothing at work, employees in support functions (such as secretaries) were often described as being the most likely to wear summer outfits. Since these individuals occupied similar indoor climatic conditions to those enjoyed by my respondents, this variation was not due to the existence of different ambient temperatures. Rather, my interviewees attributed this difference of approach to the fact that support workers were not expected to look as 'professional', since they had less need to impress clients. Others suggested that support staff had more time to think about these things. These stratified patterns of dress were then reflected back in the lawyers' views. For some, seasonally *unchanging* dress was becoming a mark of distinction. In other words, not adjusting to the seasons was taken to show that you were too busy, ambitious and preoccupied with work to notice the weather. In this way, being seen to respond to the arrival of warmer weather might suggest that you were less than fully focused at work and, by implication, less than fully committed. As such, the observed differences of practice between support staff and lawyers drew my professional respondents deeper into a seasonally unchanging community of practice.

Consistency within this group was as much a consequence of how the above dynamics combined as of any external pressure or compulsion. Vague senses of what bosses expected in terms of dress contributed to some respondents wearing the same sorts of outfits all year round. Some thought it better to stick with the usual uniform than risk the rarely spoken disapproval of superiors. Yet, many more said that there was little to worry about on that score, so long as they were not excessive and so long as a professional appearance was maintained, particularly in meetings. In other words, the lawyers with whom I spoke probably could have dressed and acted slightly differently during the summer. Indeed, when an individual did take such an approach, other respondents viewed this as a pleasing distraction, some also feeling that such seasonal changes were 'natural' and appropriate. However, despite all this, a community of shared and seasonally unchanging practice was in some ways strengthening, sustained by social cues, preoccupations with working life and career, ideas about how free time should be spent, and beliefs about the value of appearing busy.

Returning to the communities of practice literature, Wenger writes about 'brokering' (1998: 109). In his terms, brokering represents a means of introducing new ideas and variants, whilst remaining sensitive to the community and the importance that repeated performance of practice has for its continued existence. Brokering is commonly done by newer recruits who have yet to be fully absorbed into the practice, or those who must straddle the community boundary such as line managers tasked with implementing the suggestions of superiors. In theory, there may be scope for some sustainability 'brokering' before the community of practice – in this case one defined by reliance on energy intensive air conditioning, and which does not adapt to summer – solidifies further. One practical recommendation, arising from my research, was to suggest emailing employees at the start of summer, to remind them that they could wear lighter clothing during this period. This was already being done in some of the offices I studied, and was universally welcomed by my respondents.

Of course, we might imagine that communities of shared practice are readily established in office settings where members of staff have many opportunities to observe and emulate the practices of others. Yet it is worth remembering that the respondents I interviewed were drawn from different organisations. Clients and colleagues from other firms were generally ushered into formal meeting spaces, and these formal 'public' meetings were characterised by conventions of their own. In contrast, the normal desk, and the ordinary working environment, was a relatively private place. In these settings, clothing habits were as much about how people invested time and energy as about norms of public presentation. In other words, the picture was more complicated, and practice communities were less straightforwardly determined by repeated exposure to the actions of others. It is with this thought in mind that I turn to my second case.

Held outside a possible community of practice: unaware of how others manage winter cold at home

The second research project examined the ways in which a small sample of older people managed to keep warm during the winter. In the context of the present discussion, I was interested in whether these respondents' strategies were similar and whether they also constituted a community of practice. Whilst some older people find it difficult to get through the winter without hardship, others may have more profligate habits of 'wasteful' home heating. At both of these extremes, there are good reasons to encourage older people to adopt less energy intensive and less expensive methods of keeping warm. With these issues in mind, a second interview-based study was designed to discover whether variations in income, home ownership and housing type were associated with differences in how older people kept warm in winter (for more on this, see Hitchings and Day 2011). The study showed that there were, indeed, diverse ways of keeping warm at home – perhaps not surprising given the structured diversity in the sample. However, this diversity was not simply attributable to differences of circumstance, for instance in terms of wealth or housing. As we discovered, other dynamics were at work, meaning that our respondents did not form a 'community' of practitioners defined by shared strategies of winter warmth.

On being asked about what they thought older people did to stay warm at home, respondents were often eager to distance themselves from the idea that there might be a common generational response. The 'elderly' community was certainly not one with which our respondents readily identified in this regard. This was largely the product of a persistently negative media stereotype, in which older people were viewed as being passive and incapable of managing or coping with seasonal cold. This representation is consistent with images deployed by charities seeking support and by broadcasters seeking headlines. It is certainly true that some older people do face problems in the winter. However, one indirect consequence of the media stereotype of the elderly was that respondents were at pains to dissociate their own actions from those imagined or expected of others of their generation. The generational peers they saw in the media clearly could not cope, and this was quite unlike them. Rather, what individual respondents did to keep warm at home was routinely defended as the most 'sensible' strategy, given their personal circumstances. Accordingly, respondents often talked about 'doing what is right for me' – a formulation which emphasises the distance between their own responses and those they imagined might be typical of their generation.

It was rare for respondents to talk about home heating with friends and neighbours. This was thought to be a sensitive topic, particularly amongst those of a similar age. As described, raising this subject might imply that you considered the person with whom you were talking to be incapable in this respect. Our interviewees were well aware of the various declines associated with aging and felt that others would be similarly sensitive. Rather more cheerful topics, associated with family news or holiday plans, were generally taken to be safer options when talking with other older people. In this context, it was difficult to develop any collective understanding of thermal comfort practices. Consequently, there was much speculation about who did what and why. When we discussed the winter behaviours of others, the conversation frequently turned to gossip as we raked over the various reasons why others of a similar age might, or might not, have adopted certain strategies. The key point here is that knowledge was limited. Even very good friends knew little about how each achieved thermal comfort at home during the winter.

This lack of knowing was reinforced by the fact that it was considered socially important to provide a warm home when others visited and impolite to talk about being cold when visiting others. The majority of our respondents were well aware of the pleasures and benefits of a warmer ambient environment as they aged and, reflecting this shared convention, many made the house warmer in preparation for guests, and especially for guests of a similar age. There was general agreement that a host would feel ashamed if visitors kept their coats on. In addition, several respondents felt that they should wear smarter, often less insulating, clothes when others came to visit. Certainly, they did not use daytime blankets when hosting guests, though several respondents did so when at home alone. This was a distinctly private arrangement, not one to be shared or widely discussed for fear of reproducing the stigmatised stereotype mentioned above.

Social etiquette also meant that when visiting others, respondents would be unlikely to ask their hosts to turn the heating up. To do so would seem to be impolite, if not insulting. Some wore thicker items of clothing when visiting others, in order to avoid this problem, or would try to stay in warmer parts of the host's home, such as the kitchen. Again the point is that these social imperatives prevented people from knowing, in detail, about other ways of keeping warm. Since hosts provided atypical conditions for visitors and since respondents were reluctant to mention their own preferences when visiting friends, routine strategies remained private. In sum, our interviewees neither identified greatly with their generational peers, nor knew about how others of a similar age were coping with winter at home.

In this second case, variations in practice were largely invisible: hidden from those involved. There were good reasons for this invisibility, and for the lack of any recognisable community of shared practice. First, as we have seen, respondents were keen to distance themselves from the idea of having practices in common because of their age. Second, respondents were reluctant to discuss strategies of keeping warm with friends of a similar age because of the social pitfalls associated with this potentially sensitive topic. Third, they could not observe each others' practices: when hosting guests, thermostats were often raised, different clothing was worn, and blankets were put away. Likewise, when visiting others, personal preferences went unmentioned. The development of a community of practice – with respect to the management of indoor climates during winter – was impeded by all of these considerations.

Notwithstanding the diverse situations in which these older respondents lived, rather more discussion and dialogue might have helped members of this group learn from one another about effective and sustainable ways to manage the winter cold. In this context, a 'community of practice' might facilitate a more positive sense of connection and commonality and might help diffuse less resource intensive, and less expensive, alternatives to some of the current winter warmth strategies described to us in the project. Based on this understanding, one project recommendation was to find ways of helping older people to talk more freely about this otherwise delicate topic. Once again, it seemed that some form of sustainability brokering could be beneficial.

Of course, we might have guessed it would be particularly difficult to establish a shared community of domestic practice, because homes are private spaces to which others have limited access and which also take a variety of physical forms. However, this study suggested that limited sharing of knowledge and practice was due to much more than architectural difference or frequency of exposure to 'normal' or real indoor conditions within other peoples' homes. As such, this second example underlines the need for detailed analysis of the subtleties of practice community formation: it is not a matter of simply presuming that certain factors will be more significant than others. In this case, the formation of a community of practice, ordered around similar modes of achieving thermal comfort at home, was impeded by various social conventions in addition to differences of wealth, tenure, and housing type. Avoiding the stigma of being 'old' and managing related aspects of social nicety appeared to be critical considerations, and both were important in preventing the development of a shared practice.

Two kinds of insulation

In this chapter, I have taken a comparatively unusual approach to the discussion and analysis of sustainable communities. Rather than defining these in terms of neighbourhood or geographical proximity – 'communities of place' – I have identified 'communities of practice' that might be targeted and engaged in pursuit of a more sustainable society. I then considered how social networks and conventions, along with more formal organisational arrangements, structure the knowledge people have of each others' practices, and the extent to which this knowledge is shared. My method was to identify specific features of two social settings that encouraged and/or discouraged the formation of communities of 'thermal practitioners', by which I mean communities united by the way in which they relate to indoor environments, and by their approach to keeping warm or cool.

This led me to suggest that deliberate intervention, in the form of sustainability 'brokering', might be useful as a means of encouraging the emergence of new communities, organised around variants of thermal practice, which are less energy intensive to maintain. By paying close attention to the specific dynamics involved, both projects generated ideas about how this might be done. In the first case, the proposition was to actively promote the idea of wearing lighter clothes in summer. Intervention of this kind might be effective in that changing into a summer outfit is already socially acceptable: respondents were not against the idea – it was just not something they thought about. In this situation, where a 'community' is already formed, it might be possible to redefine the practice that is shared. For example, the community that is currently defined by an expectation of mechanical cooling and a disregard for the seasons might come to share new conventions, including those of seasonal adaptation. In the second case, the rather different challenge was that of helping to bring a community of shared practice into being. Here, the practical suggestion was to develop and promote situations in which older people might share ideas and talk together about effective and efficient methods of keeping warm during the winter months. Our research indicated that respondents were eager to know about, and potentially learn from, their generational peers.

The two studies generated rather specific ideas about what might be done to influence practice community dynamics in each context. Are there other more broadly relevant conclusions to take from this exercise? In conclusion I focus on two. Both are about insulation, albeit of different kinds.

The first kind of 'insulation' is physical. Most discussions of energy demand focus on the details of home heating or air conditioning and the resources and technologies required to control and maintain ambient conditions. If we concentrate, instead, on the practices involved in keeping warm or cool indoors, it becomes obvious that this is only part of the story. What people wear, and how concepts of appropriate clothing vary from one social situation to the next, is also crucial. This is not a topic that figures prominently in energy research, but it is one that clearly deserves more attention. In this respect, a 'communities of practice' approach could help in understanding how conventions and practices that matter for thermal comfort circulate between people.

Asking about when people put blankets away, what they mean by casual and formal dress, and what they wear in different contexts might seem trivial, even banal. However, these questions are important if we are to understand how more (and less) resource intensive practices circulate, and how understanding of such processes might be channelled to positive environmental ends. As these two small studies have demonstrated, there is considerable and persistent variation in how thermal comfort is achieved and in the cultural and physical contexts through which related practices are, and are not, shared. Attending to the community dynamics involved could help identify sensitive methods of encouraging lower carbon variants to flourish and others to decline.

The second kind of insulation is social. When used in this way, the term refers to the possibility that people might be insulated from the knowledge or sense that other strategies might be possible. In effect, social networks insulate people in different ways. This is normal and not especially problematic. After all, insulation of this kind allows people to go on with daily life without thinking about whether they are doing the right thing and without being troubled by the sense that there are other possible options. At the same time, patterns of social insulation may be decidedly unhelpful for those hoping to disrupt current practices in pursuit of environmental ends. This argues for better understanding of how people are held together inside communities of practice, and of how such communities do and do not form. My examples suggest that the processes involved are more complex than one might at first expect, and that it is important to examine detailed differences (and similarities) in how practices are reproduced and shared. Finally, I have suggested that knowing more about how communities of ordinary practice develop can be of immediate practical value for those seeking to dislodge resource intensive arrangements and promote new communities, ordered around more sustainable ways of living.

Acknowledgements

The two projects discussed in the paper were funded by the ESRC (Ref: RES-000–22–2129-A) and the Nuffield Foundation (Ref: SGS/36294). The second was in collaboration with Rosie Day (Geography, Birmingham University).

References

Aiken, G. (2012) 'Community transitions to low carbon futures in the Transition Town Network', *Geography Compass*, 6 (2): 89–99.

Chappells, H., Medd, W. and Shove, E. (2011) 'Disruption and change: drought and the inconspicuous dynamics of garden lives', *Social and Cultural Geography*, 12 (7): 701–715.

Connors, P. and McDonald, P. (2011) 'Transitioning communities: community, participation and the Transition Town movement', *Community Development Journal*, 46 (4): 558–572.

Evans, D. (2011) 'Consuming conventions: sustainable consumption, ecological citizenship and the worlds of worth', *Journal of Rural Studies*, 27: 109–115.

Franklin, A., Newton, J., Middleton, J. and Marsden, T. (2011) 'Reconnecting skills for sustainable communities with everyday life', *Environment and Planning A*, 43 (2): 347–362.

Hitchings, R. (2011) 'Researching air-conditioning addiction and ways of puncturing practice: professional office workers and the decision to go outside', *Environment and Planning A*, 43 (12): 2838–2856.

Hitchings, R. and Day, R. (2011) 'How older people relate to the private winter warmth practices of their peers and why we should be interested', *Environment and Planning A*, 43 (10): 2457–2467.

Lave, J. (1993) 'The practice of learning' in S. Chaiklin and J. Lave (eds) *Understanding Practice: Perspectives on Activity and Context*, Cambridge: Cambridge University Press.

Lave, J. and Wenger, E. (1991) *Situated Learning: Legitimate Peripheral Participation*, Cambridge: Cambridge University Press.

Reckwitz, A. (2002) 'Towards a theory of social practices: a development in culturalist theorizing', *European Journal of Social Theory*, 5 (2) 243–263.

Seyfang, G. (2008) *The New Economics of Sustainable Consumption: Seeds of Change*, London: Palgrave.

Seyfang, G. and Smith, A. (2007) 'Grassroots innovation for sustainable development: towards a new research and policy agenda', *Environmental Politics*, 16 (4): 584–603.

Shove, E. (2003) *Comfort, Cleanliness and Convenience: The Social Organisation of Normality*, Oxford: Berg.

Shove, E. (2010) 'Beyond the ABC: climate change policy and theories of social change', *Environment and Planning A*, 42 (6): 1273–1285.

Turner, S. (1994) *The Social Theory of Practices*, Chicago: University of Chicago Press.

Walker, G., Hunter, S., Devine-Wright, P., Evans, B. and Fay, H. (2007) 'Harnessing community energies: explaining and evaluating community-based localism in renewable energy policy in the UK', *Global Environmental Politics*, 7: 64–82.

Wenger, E. (1998) *Communities of Practice: Learning, Meaning, and Identity*, Cambridge: Cambridge University Press.

Part IV
Relations between practices

8 Building future systems of velomobility

Matt Watson

Introduction: system transition and practices

What would it take for cycling to displace driving as the dominant practice of personal mobility?

The challenge of bringing about change on such a scale appears to be beyond the realm of possibility. In this chapter, I argue that it is nonetheless useful to think about the means of achieving, or at least working towards, such an improbable future scenario. It is well established that reducing car use and increasing cycling has many potential benefits, including improving individual health and wellbeing, mitigating the effects of cars on public space and air quality and limiting greenhouse gas emissions from car exhausts (Michaelowa and Dransfeld 2008; Woodcock *et al.* 2009). In exploring the question of whether cycling might displace driving, my purpose is not to proselytise for cycling, but to consider the relevance of theories of practice for illuminating future systemic shifts towards less carbon intensive modes of organising daily life.

It is possible to trace the co-evolution of practices and the broader sociotechnical systems of which they are a part, for example in relation to doing the laundry or maintaining thermal comfort (Shove 2003). In theory, retrospective analyses of how technical innovations, changes in product design or developments in infrastructure shape the directions in which practices develop, provide useful clues for those interested in steering such trajectories in line with normative commitments (Shove *et al.* 2007; Watson and Shove 2008). By implication, those wanting to engender systemic transitions would do well to begin by thinking about how practices are sustained by, and themselves sustain, sociotechnical systems. We know that transitions occur because of the dynamic interrelations between technologies, infrastructures, markets, norms and regulations that constitute socio-technical systems (Rip and Kemp 1998; Geels *et al.* 2004; Elzen and Wieczorek 2005). We also know that transitions, whether in the past or in a hoped for future, cannot be reduced to individual choice and behaviour. And yet, they only come about if enough people do enough things differently enough. So, how do we link changes in what people do to the complex, heterogeneous socio-technical arrangements, which must be radically reconfigured if we are to militate meaningfully against climate change?

For those conversant with theories of practice, the move is obvious. What people do is never reducible to attitudes or choices, or indeed, to anything simply individual. Rather, doing something is always a *performance* of a *practice*. From a practice perspective, enough people doing enough things differently enough is not a matter of atomised individuals choosing to act in another way. It is a matter of the dynamics of practice (Shove *et al.* 2012). Practices, and socio-technical systems, are mutually constitutive in that it is through practices that socio-technical systems become embedded in the routines and rhythms of daily life. At the same time, the everyday performances that reproduce (and transform) practices are shaped by the socio-technical systems of which they are a part. System transitions only happen if the practices which anchor those systems in daily life change; likewise if practices change, then the socio-technical system must also evolve (at least incrementally).

In what follows, I work through the relevance of these ideas by considering what it would take for the practice of cycling to recruit practitioners and colonise travelling to the point of displacing driving. In the first section, I discuss the contested relation between co-existing practices of cycling and driving and suggest that this can be conceptualised with reference to two distinguishable systems: one of velomobility, the other of automobility. These two systems, while morphologically similar and mutually interdependent, also compete with one another. I argue that this competitive aspect is crucial for understanding and characterising the dynamics of personal mobility. I then focus on the specific practices of cycling and driving, and on processes of recruitment and defection through which 'carriers' are enlisted and lost. I contend that an emphasis on systemic transitions in practice can help specify opportunities for intervening with the aim of reconfiguring socio-technical systems of personal mobility. The final section of the chapter provides a worked example of what this might involve. In conclusion, I consider recent initiatives to promote cycling, using these cases to identify some of the transformations that would have to occur in order to achieve a major transition towards systems and practices of velomobility.

Thinking with cycling and driving

There are various reasons why thinking about a shift from auto- to velomobility might be relevant for understanding the relation between practices and socio-technical systems. First, it would not be easy to represent such a transition in terms of any conventional narrative of innovation. The present situation is one in which bicycles, which have been around for longer, have generally lost out in competition with cars. If there is to be any systemic transition towards velomobility, it will have to emerge not so much from niches of innovation as from pockets of persistence, these being situations in which people have carried on riding bicycles in the systemic interstices of automobility.

Second, despite there being an element of competition, bicycles do not simply substitute for cars. Practices of car-driving and bicycle-riding currently fit into the spatial and temporal requirements, and purposes, of practitioners' lives in very different ways. Even so, the possibility of figuring system transition as a struggle between competing systems, rather than a process of evolution within a single system, offers distinctive intellectual challenges and is itself useful in articulating and developing the analytical potential of a 'practice orientation' for conceptualising systemic transitions.

Third, there is increasingly detailed evidence of wide variation in the relative dominance of velomobility in different places and times. There are also local examples in which deliberate efforts to promote cycling have the desired effect. This means that there is a body of relevant data documenting the effects of systemic interventions on the dynamics of cycling practice. These data can be used to inform otherwise speculative future-gazing.

Finally, the changing relation between velo- and automobility has direct implications for thinking about the relations between transitions in practice and climate change. Cars are icons of unsustainable modernity. Cars consume more than 10 per cent of total primary energy supplied in the UK (DECC 2009), and transport stands out as the only sector to see an increase in emissions – of 48 per cent – between 1990 and 2003 (Office of National Statistics 2004). Globally, personal mobility accounts for around 26 per cent of anthropogenic CO_2 emissions (WWF 2008).

Of course, the claim that cycling is 'the zero emission option' is wrong. Even cyclists, reckoned to be the most efficient moving animal on land (Wilson 1973), cannot altogether escape the laws of thermodynamics. A typical cyclist uses around 80 kJ/pkm[1] of food energy in excess to resting metabolism – less than one-thirtieth of that required by a car with one occupant (MacKay 2009). However, as systems thinkers, we know we need to consider the embodied energy of the fuels involved. Coley (2002) calculates the embodied energy of food (taking into account the energy inputs along the supply chain) typically used by cyclists to be 539 kJ/pkm. This appears to dent the benefits of bicycles over cars until we consider the possible inefficiencies of car fuel production. If we take account of embodied energy, the 'cost' of petrol and diesel (in terms kJ/pkm) increases by around 40 per cent (MacKay 2009). Although already becoming rather complicated, our calculation has yet to consider the embodied energy of the vehicles involved (the embodied energy of bicycles is estimated at 190 kJ/km; cars around 1830 kJ/km (Lovelace *et al.* 2011)). What then of the relative costs of roadways for cyclists and cars? How do we account for cyclists' use of infrastructures engineered for HGVs, whilst knowing full well that bicycles could spin along fine on much less energy intensive surfaces? What if cyclists take more showers and launder more clothes, as well as eat more food, as a result of their sweaty exertions? How might this be offset against fewer trips to the gym? And so we could go on. The already complex project of contemplating the energy and greenhouse gas implications of the balance between car-driving and bicycle-riding, is even more challenging if we consider arguments about the wider societal implications of a shift towards velomobility in terms of individual health, social inclusion and cohesion (Komanoff 2004; Woodcock *et al.* 2007).

120   M. Watson

Notwithstanding the uncertainties that crowd in around efforts to estimate energy, carbon and other properties, it is clear that in the face of peak oil and climate change, a transition from automobility to velomobility would be a good thing. It is also clear that purposive intervention is required if such a transition is to be effected. But what does it mean to think of velomobility as a complex socio-technical system? And what might intervention involve?

Systems of auto- and velomobility

Geels (2004) uses the car as an example with which to introduce the idea of a socio-technical system. In his account, the vehicle is only one component in a socio-technical arrangement that includes roads and traffic systems, along with infrastructures of fuel production and distribution (from oil wells to petrol stations), and of auto manufacture, maintenance and sales. This 'system' also encompasses regulation and policy, market structures, cultural and symbolic meanings and user practices (Geels 2004: 20, 2005). Representing these elements as part of a system is not simply to place them together in a rag-bag of component parts. Rather, the point is to understand how these diverse elements inter-relate in structured and systemic ways and to identify *processes* that lead to the emergence of a particular dominant structure of personal mobility (Shove 1998).

Urry (2004) draws these elements, and more, into his exposition of the 'system of automobility', highlighting the complex interdependencies and feedback mechanisms that link technology, infrastructures, markets and meanings, and that converge around the hybrid entity of the car-driver. For Urry:

> [a]utomobility can be conceptualized as a self-organizing autopoietic, non-linear system that spreads world-wide, and includes cars, car-drivers, roads, petroleum supplies and many novel objects, technologies and signs. The system generates the preconditions for its own self-expansion.
>
> (Urry 2004: 27)

By conceptualising the car's rise to dominance in terms of its positioning within a complex emergent system, Urry describes what he refers to as a process of systemic self-extension. At the end of the nineteenth century, a series of contingent turnings and small causes in the development of motorised transport seemed to develop, irreversibly, into a socio-technical lock-in, organised around the petrol and steel car. The system about which Urry writes becomes progressively extended and embedded in car technologies and in the activities and economic interests of oil companies. More profoundly, this system extends to the restructuring of space, through the growth of suburbs independent of train lines, and through the progressive appropriation of land for roads and for parking, especially within cities. As Urry explains, the re-making of urban environments around the car entails a restructuring of temporalities: the car enables the fragmentation and speeding up of tasks distributed across space. Through the complex interaction of these different elements, the car arguably creates the conditions of its own necessity and the means for its own extension. So it is, that automobility comes to exert its very decisive 'character of domination' (Heidegger, in Sheller and Urry 2000: 737), becoming 'one of the principal socio-technical institutions through which modernity is organized' (Böhm *et al.* 2006: 3).

To some extent, the emergence of a system of velomobility could be described in very similar terms. A range of elements, relationships and actors also crowd around the technological artefact of the bicycle (or the hybrid entity of the bicycle-rider). Systems of velomobility have road traffic infrastructures and associated institutions, often shared with the car. Processes of manufacture, distribution, maintenance and repair are important, and cycling is also regulated and governed. Likewise, bicycle production is shaped by global markets. Cycling, like driving, represents a means of moving human bodies from one place to another and, like the car-driver, the cyclist has to possess distinctive skills and bodily competencies. Discourses and representations circulate around the bicycle as they do around the car and although the experience differs, both generate a distinctive phenomenological engagement with the world (Spinney 2007). In brief, systems of velo- and automobility have similar composition and conceptual shape. So how has automobility acquired this quality of relentless self-extension, while velomobility struggles to persist alongside the more dominant system? Could the system of velomobility arrive, by similar means, at a tipping point – a threshold beyond which it becomes, at least locally, *the* dominant form?

In different times and spaces, velomobility has had just such a character of systemic self-extension. Histories of the emergence of cycling as a means of mass transit show that it grew very rapidly during the late nineteenth century in many industrialised countries. From this point of view, the bicycle's history maps neatly on to established models of innovation, normalisation and socio-technical change (Geels 2002). There was, for instance, a long period of iterative technological innovation during which bicycle-riding was largely restricted to wealthy young males, its purposes more defined by enjoyment and risk taking than transport utility. However, the stabilisation of the bicycle as technological artefact (Bijker 1997) underpinned the broader stabilisation of an emergent socio-technical system. This provided the basis for the wider diffusion or 'breakthrough' of cycling in the early twentieth century. While details of timing and the extent of appropriation varied across northern Europe, in many countries rates of cycling rose to a peak at some point during 1940s. Based on survey data and interviews in Britain, Pooley and Turnbull (2000) conclude that in the 1930s and 1940s, around one-fifth of men and one-tenth of women cycled to work. They go on to explain that in smaller settlements cycling was, for a time, *the* single most important method of travelling between home and work. The machinic complex of the bicycle has certainly seen at least local dominance before, and in some parts of the world it retains this status. For example, the modal share of cycling in Tianjin, China, was recently reckoned to be 77 per cent (International Bicycle Fund nd).

In the UK in 1952, cycling accounted for some 23 billion km, or 13 per cent modal share (DfT 2006). However, this fell dramatically and, by 1972, cycling had declined to 5 billion km, or just 1 per cent modal share (DfT 2006). Distance and modal share have stayed roughly stable ever since (Cabinet Office 2009). This radical decline of cycling, and its repositioning as a form of recreation rather than a utilitarian means of transport, was mirrored to a greater or lesser extent across northern European nations. Not surprisingly, this transformation coincides with the rise of the car as an increasingly democratic means of personal mobility. Reasons why cars might take the place of bicycles are legion, including issues of speed, safety, comfort and status.

However, the decline of cycling, and the concurrent rise of driving, is not a tale of direct technological substitution (Shove 2012). The story is not simply one in which people buy a car and then hang their bicycle in the shed. For a start, there are key areas of systemic symbiosis. First, the growth of the system of velomobility provided some of the elements incorporated in the coming system of automobility (Geels 2005). These included aspects of production capacity and infrastructure, along with ideals and meanings of mechanised personal mobility. Second, cars do not simply replace bicycles, but afford a different range of uses, meanings and purposes.

If we are interested in promoting systemic transition away from automobility, it is worth paying close attention to the web of interdependent relations that developed alongside, and as part of, the growing dominance of the car. In the next section, I consider cycling and driving (and the related practices on which both depend) doing so as a method of interrogating the relationship between changes in what people do and the dynamics of socio-technical systems.

Theories of practice do not take the individual subject as the centre of enquiry, but in drawing attention to the changing performances of cycling and driving as enacted by individuals, they recognise that drivers and riders play a crucial part in making systems of velo- and automobility. This role is typically understated within dominant narratives of socio-technical change. In Urry's account, the system of automobility is one in which individuals are coerced by the increasingly self-evident necessity of car transport. However, as highlighted in the introduction to this chapter, it is clear that systems can only emerge, persist and gain dominance by capturing time and attention, and by colonising what people do. Whilst theories of systemic transformation challenge simplistic models in which individual attitudes are thought to drive behaviour, they tend to overlook the practicalities and details of doing. In other words, understanding transitions in socio-technical systems depends on paying attention to the trajectories of *practices* that are themselves active constituents of those systems. In the next section, I explore this proposition by considering the shifting relation between velo- and automobility with reference to changing patterns of recruitment to, and defection from, bicycle-riding and car-driving (or passengering).

Practices of cycling and driving

Cycling and driving can each be understood as a practice (or coherent bundle of practices). As Schatzki suggests, a practice exists at once as an entity, a 'nexus of doings and sayings' (1996: 89) and as a performance. Accordingly, '[p]ractice in the sense of do-ing ... actualizes and sustains practice in the sense of nexuses of doings' (1996: 90). A practice, like cycling, exists as an *entity*, enduring beyond and between individual performances or enactments. As an *entity*, cycling exists as something that can be spoken of and about. There are, in addition, many material traces (bicycles, accessories, road signs, bike shops etc.) which persist between moments of performance. However, cycling only exists as an *entity* because it is performed, most obviously through people riding bicycles. These 'doings' are shaped by the complex relations, between knowledges, norms and materials, which form the practice as an entity.

From this point of view, it is clear that practices are themselves *part* of socio-technical systems. For example, driving is entirely dependent upon the technologies and infrastructures that make the practice possible, and these are, in turn, dependent upon global flows of oil, steel, meaning and money. Driving is enabled and sustained by institutions which regulate, legislate and enforce. In short, the practice of driving cannot exist, at least not in anything like its present form, without most of the elements of this wider system of automobility as currently configured.

However, the reverse of this is equally true. That is, the relations constituting the present system of automobility would not exist without the continued performance of the practice of driving. Performances of driving do not only serve to reproduce the practice of driving as an entity. They also reproduce the complex systemic relations with which the practice of driving is co-dependent. Moreover, as performances of driving become more frequent and undertaken by more practitioners, they demand and engender the *extension* of those systemic relations. As Shove and Pantzar (2005) argue in relation to Nordic Walking, the processes by which practices *recruit* practitioners are inseparable from, and co-constituted with, processes of innovation in relation to technologies, knowledges and meanings. With Nordic Walking, as with the more complex system of automobility, processes of recruitment are central: understanding how people become drivers or cyclists is consequently vital in understanding how these systems extend and grow.

As indicated above, the rise of automobility cannot be separated from the decline of velomobility. The increasing domination of the car is thus as much a story of defection from cycling as of recruitment to driving. As practices, cycling and driving compete for many of the same resources. Like all practices, they compete for practitioners' *time*. This competition is more direct when performances of the practice might fall into the same slots of temporal routine, or relate to the same project of moving from one place to another. They also compete for *space* on roads and in cities. They compete for *money* in complex ways – once the major investment of a car is sat on the driveway, and depreciating, there might be a tendency to use it even when cycling is an obvious and viable alternative. Finally, they compete in discursive and symbolic realms, and in debates about safety, health, responsibility, convenience and status. In brief, the widespread decline in cycling from the middle of the twentieth century onwards can be understood as result of automobility 'winning' across a range of systemic level competitions.

From this point of view, practices, and the processes through which they recruit or lose practitioners, constitute the motor of systemic obduracy and change. They do so in the sense that practices (as entities), and the socio-technical systems of which they are a part, are embedded, reproduced and iteratively reshaped through successive moments of performance. It follows that a future shift from auto- to velomobility depends on increasing recruitment to cycling and defection from driving. This is not a matter of somehow rewinding processes that led to the dominance of automobility in the 1950s. The relative significance of cycling in the 1940s was part of a different socio-technical landscape and the ways in which fast, heavy, hard motor vehicles displaced bicycles, both physically and semiotically, cannot be reversed. Cycling has been shaped, over the last few decades, by its subordination to the car. Looking ahead, the challenge is therefore one of identifying opportunities for increasing recruitment to velomobility within the current socio-technical landscape. The discussion so far suggests how demanding it would be to engender a shift from auto- to velomobility. The former system has established enormous competitive advantage in recruiting practitioners and sustaining performances over many years and has become ever more locked-in to infrastructures, political priorities, spatial planning and organisation and experiences of time and space.

Can we, nonetheless, imagine possible steps on an inevitably iterative path towards the future dominance of velomobility? In the remainder of this chapter, I suggest that if we are to pursue this idea, we need to focus on processes of recruitment and the dynamics of practice. This approach provides a way of framing the problem that is distinctive on at least three counts. First, what are conventionally seen as barriers to participation in cycling (usually attributed to individual attitudes) are usefully reconceptualised as 'sticking points', that emerge from systemic relations between cycling and driving. Second, it is evident that whether intended or not, policy interventions are interventions in the dynamics of practice. Third, and following arguments made above about the relations between the dynamics of practices and transitions in systems, effective interventions are likely to be those which initiate or add momentum to processes of positive feedback in patterns of recruitment to cycling, or defection from driving. The following sections consider each of these three features in turn.

Systemic sticking points

Studies of individual attitudes and enquiries into why people do, and do not, choose to cycle, reveal a perennial range of 'barriers'. These seemingly personal obstacles can be reconceptualised as 'systemic sticking points': that is, as effects which emerge from particular combinations of systemic relations, and that hamper recruitment to the practice of cycling, or promote defection from it.

One of the most prominent is the issue of safety. Just how high the risks of cycling appear depends on how the statistics are presented. In 2005, one cyclist was killed for every 30 million km travelled by bike in the UK (CTC nd). Cycling is twice as safe as motorcycling per km travelled, and you are only a little more likely to get killed or seriously injured cycling one kilometre than you are walking it (Hamilton and Rollin Stott 2004). However, the appeal of cycling diminishes in the face of figures showing that it is 20 times safer to travel that one kilometre in a car than on a bike.

Time and convenience represent other commonly cited barriers, often leading people to conclude that it takes too long, or takes too much effort, to cycle the distances that separate home and work. Issues of weather and topography are also important for the prospect and the experience of travelling by bike. Further sticking points, often less obvious in quantitative studies, relate to issues of physical effort and fitness; sartorial conventions and norms of personal hygiene (lycra, fluorescent jackets, sweat, etc.); the risk of bicycle theft and so on.

Each of these sticking points can be positioned not as the idiosyncratic response of individual travellers, but as culturally distributed, profoundly contingent, properties of the social. These representations of cycling are contingent in that, in other times and spaces, the relation of cycling to distance and time was much more positive. It is only in competition with cars, and only in a world reshaped around cars, that distance becomes a sticking point rather than an inducement to cycling. It is only when daily routines are stretched between spaces and compressed in time, and only when cities are reshaped around the car that cycling has difficulty competing. Similarly, concern around safety issues develops when cyclists are forced to occupy the margins of roads and routes built around the car (92 per cent of cycling fatalities result from collision with a motor vehicle (Hamilton and Rollin Stott 2004)), and when car-driving transforms societal expectations and experiences of physical risk. Anxieties about unattractive clothing and smelliness are no less systemically embedded, even if they are softer, more malleable and perhaps less critical (Shove 2003). In sum, what look like barriers preventing individuals from getting on their bikes are better understood as systemic sticking points in longer term processes of recruitment to the practice of cycling. Locating them as such helps identify points of intervention that have yet to figure in the repertoire of established policy tools.

Political interventions in practice

Hard evidence for the efficacy of policy measures is patchy (Krizek *et al.* 2009). However, piecemeal attempts to create cycle lanes at the edges of roads, construct short off-road sections and install more secure cycle racks are unlikely to produce any substantial change in recruitment to cycling. Sure enough, in the UK, national strategies of this kind have failed to reverse the long-term decline of cycling and walking (Cabinet Office 2009).

In thinking about how policy might intervene in a more effective and more systemic manner, it is useful to compare experiences in London and Groningen, two cities in which systems of velomobility are variously well embedded.

Cycling declined in the Netherlands, as in the UK, through the 1950s–1970s, but did not fall as far (in the Netherlands rates dropped by 62 per cent between 1950 and 1975, compared with 80 per cent in the UK). Cycling has not recovered in the UK, but in the Netherlands there was a tangible resurgence from 1975 until the 1990s and rates of cycling have continued to increase such that 27 per cent of trips are by bike (the highest modal share in Europe), compared to just 1 per cent in the UK (Pucher and Buehler 2008). There are other notable differences. Striking demographic inequalities in cycling are apparent in the UK, but there are fewer in the Netherlands. In the UK, men make 72 per cent of all bicycle trips, while in the Netherlands they make only 45 per cent. In the Netherlands, Denmark and Germany, 'cyclists comprise virtually all segments of society' (Pucher and Buehler 2008: 502).

This context is important in that in Groningen today, almost 40 per cent of local trips are made by bike. This reflects long-term political commitment to cycling through fundamental, systemic priorities, executed via mutually reinforcing policies of compact land-use, instruments to restrict car-use and investment in cycling infrastructure. In Groningen, as in many cities in the Netherlands, Denmark and Germany, transport policies systematically favour cyclists at the expense of drivers.[2] Policies to promote the practice are quite unlike those adopted by towns in the UK. For example, in Groningen, there are no campaigns to persuade cyclists to wear helmets or encourage people to use the bike. There is no need for such initiatives in that the city's strategy for efficient personal transport is already shaped around the bicycle and cycling is already an entirely normal thing to do (Pucher and Buehler 2007, 2008).

Such as it is, London's system of velomobility has very different characteristics. Rates of cycling in London followed the UK's trend, declining to a low point in the early 1990s. From there cycling began to increase gradually until 2003. In that year, the introduction of the congestion charge coincided with a period of rapid growth. Rates of cycling rose by at least 50 per cent between 2003 and 2007 and continue to escalate. Of course, this was from a very low base so, while in some boroughs cycling is reported to account for 10 per cent of journeys to work (London Travel Watch 2009), across the city as a whole, the share of journey stages made by bike was around 2 per cent in 2007 (Transport for London 2007). This change in the relative dominance of cycling cannot be attributed to the congestion charge alone. Substantial investment has also been made in cycling infrastructure, yet the rate of recruitment appears to be increasing faster than could be accounted for by specific interventions such as the congestion charge or 'London Cycle Network +'; this indicates the existence of secondary and feedback effects, which may offer clues as to the circumstances in which velomobility might develop qualities of self-organisation and self-extension, potentially to the point of challenging systems of automobility.

Feedback effects on recruitment and systemic change

These two cases – Groningen and London – illustrate the importance of recruitment and defection, and point to differences in how this works at different scales (and at different moments) in the ongoing history of bicycle-based systems and their embedding and integration in practice. These cases also suggest that in certain circumstances, deliberate interventions (like congestion charging) can spark off indirect processes that have unpredictable, but sometimes cumulative, consequences. A few examples illustrate this potential.

It is now widely accepted that, in general, cycling becomes safer when more people do it (Jacobsen 2003; Komanoff 2004; Woodcock *et al.* 2007). Komanoff estimates a 'power law' relationship of approximately 0.6 between cyclist numbers and cyclist safety, such that if the number of cyclists doubles, the number of accidents per cyclist-kilometre should reduce by more than 30 per cent (Komanoff 2004: 148). More people cycling can have a bigger effect on cycle safety than end-of-pipe solutions, such as promoting or legislating for cycle helmets. This can have a positive feedback effect: as safety increases, recruitment to cycling becomes easier, which in turn makes it safer again.

More speculatively, there is another possible systemic effect in which the increasing feasibility of cycling leads to a reduction in levels of car ownership. Once cycling is sufficiently embedded in a city and in a person's life and routines, the cost of owning a car (or, more likely, the cost for a household of owning a second car) may become difficult to justify.

A more diffuse effect comes through the diversification of practice as rates of cycling increase. While there is limited scope for straightforward technical innovation, there is growing evidence for the development of niches of innovation in culture and practice around cycling. For example, the first stages of what might be cycling's systemic self-extension in London have coincided with the emergence of multiple cycling sub-cultures, such as those around fixed-gear bikes, or 'velo-chic', the followers of which resist associations between the bicycle and fluorescent/lycra gear and instead adopt a style of conspicuous elegance. Specialist shops are now importing cargo bikes or box-fronted tricycles from northern Europe. Although these styles are (now) decidedly foreign to the UK, they help in overcoming one of the key obstacles to cycling: namely the challenge of carrying stuff (or several children). A proliferation of variants and versions of the practice of cycling increases possible points of contact through which new practitioners can be recruited.

These are intriguing trends, but more is required if cycling is to become the dominant mode of personal mobility. For a start, feedback processes of the kind discussed above would have to spill out across the 'system' as a whole, and in so doing, transform existing relations between bicycles and cars, along with the infrastructures on which both depend. There are different aspects to consider when thinking about how this might occur. First, recruitment to cycling necessarily means defection from other practices including (to some extent) driving. Second, as patterns of mobility shift, the demands upon the road infrastructure change such that, at a certain level of successful recruitment to cycling, priorities for investment in roads might also change, perhaps to the point of favouring cyclists over drivers (as is already the case in some Dutch cities). Third, the formal and tacit rules of the road would be re-written – Komanoff's power law implies that the details of how driving is performed are affected by incremental increases cycling, indicating that adjacent practices adjust in relation to each other. Fourth, discourses and meanings would inevitably change. The local dominance of velomobility would necessarily entail cycling becoming a thoroughly ordinary thing to do. In such a context, special associations – for example, those with lycra and fluorescent outfits, with fixies or with velo-chic style – might persist, but as niches of peculiarity within a practice of cycling that has become pressingly mundane. Expectations and norms of physical exertion (and perhaps sweat) would also change as cycling took hold.

These dynamic processes would play out in concert, one mutually reinforcing another. However, they would only enhance the overall system of velomobility when accompanied by increasing recruitment to cycling. The transition path would not be smooth, and we might expect threshold effects and turning points. For example, one such moment might arise when cycling has become so common as a mode of urban transport that formal rules of the road have to be rewritten to make cycling safer and faster and to reflect its priority over driving. If these processes of transition were to persist, political discourses would eventually revolve around cycling (not driving). There would be far-reaching consequences for the economy and industry as the material requirements and markets of cycling displaced those of driving, changing the fortunes of retailers, manufacturers and the supply chains of which they are a part.

All these speculative changes depend on patterns of recruitment to cycling and defection from driving. As indicated above, such trends imply and drive corresponding forms of socio-technical transition. For example, a radical reordering of auto- and velomoblity would require new practices of road design, building and maintenance; of legislation, governing and policing, and of manufacturing and retailing. The elements, relations and properties comprising socio-technical systems are only made, reproduced and shifted through practices enacted within and between many different sites; on the roads, in the offices of state, on the factory floor and in places where people live and work. In the end, socio-technical transitions are only effected when multiple practices co-evolve across these diverse locales.

Conclusion: building a future system of velomobility?

It would take deliberate and concerted effort to displace present systems of automobility and establish cycling as the dominant mode of personal mobility. The bicycle cannot reshape the world around itself as the car has. It is clear, from examples where cycling has become normal again, that it takes political purpose and a readiness to stand against the powerful cultural and material logics of automobility. It is also clear that getting from here to there is a matter of following (as much as steering) a complex system of practice, and that political opportunities for intervention are themselves an emergent effect of already dynamic processes.

Interventions will be more effective where policy makers recognise that the challenge is essentially one of enabling practices of velomobility to recruit practitioners, and where they recognise that this is not simply a matter of changing people's minds about cycling. This is more than a semantic shift of emphasis in that a focus on practices, and the socio-technical systems of which they are part, has practical consequences for what policy makers do and for how they view their role. For example, policy makers inspired by the ideas outlined here would not count kilometres of designated cycle paths or tot up the sums of money spent on publicity and education, as if these bore any direct relation to the resurgence of the practice. Rather, they would devote themselves to the task of identifying moments and forms of intervention that stand to make a difference, and that (within a given set of circumstances) have the potential to initiate forms of positive feedback, which favour the emergence of an entire system of velomobility. To put the point more bluntly, rearranging the details of present infrastructures will only be effective if such interventions set new patterns of recruitment and defection in train, and if these, in turn, engender system level transformation. In short, there can be no systemic transition in practices of personal mobility without parallel transitions in practices of governing, manufacturing, investing, profit making and urban living.

To conclude, the project of deliberately promoting a systemic transition towards velomobility is improbably ambitious. However, thinking about what such a project might entail has been useful in showing how relations between co-existing systems of practice might nonetheless structure transitions towards sustainability.

Notes

1 pkm = person kilometre. This unit represents the energy it takes to move one person one kilometre.
2 For example, changing priorities at traffic lights so that cyclists can travel continuously through green lights if they maintain a typical cycling speed (20 kmh); slowing and widening the turns cars make at junctions and closing areas of the city centre to cars.

References

Bijker, W. (1997) *Of Bicycles, Bakelites and Bulbs: Towards a Theory of Sociotechnical Change*, Cambridge MA: MIT Press.
Böhm, S., Jones, C., Land, C. and Paterson, M. (2006) 'Introduction: Impossibilities of automobility', *Sociological Review*, 54 (1): 1–16.
Cabinet Office (2009) *An Analysis of Urban Transport*, London: Cabinet Office.
Coley, D. (2002) 'Emission factors for human activity', *Energy Policy*, 30 (1): 3–5.
CTC (nd), Cycling Statistics: Cyclists Touring Club, available online at: http://beta.ctc.org.uk/ctc-cycling-statistics (Accessed 7.7.2012).
DECC (2009) *Digest of United Kingdom Energy Statistics*, London: Department for Energy and Climate Change.
DfT (2006) *National Travel Survey: 2006 Final Results*, London: Department for Transport.
Elzen, B. and Wieczorek, A. (2005) 'Transitions towards sustainability through system innovation', *Technological Forecasting and Social Change*, 72 (6): 651–661.

Geels, F. (2002) 'Technological transitions as evolutionary reconfiguration processes: a multi-level perspective and a case study', *Research Policy*, 31: 1257–1274.

Geels, F. (2004) 'Understanding system innovations: a critical literature review and a conceptual synthesis', in B. Elzen, F. Geels and K. Green (eds) *System Innovation and the Transition to Sustainability*, London: Edward Elgar.

Geels, F. (2005) 'The dynamics of transitions in socio-technical systems: a multi-level analysis of the transition pathway from horse-drawn carriages to automobiles (1860–1930)', *Technology Analysis & Strategic Management*, 17 (4): 445–476.

Geels, F., Elzen, B. and Green, K. (2004) 'General introduction: system innovation and transitions to sustainability', in B. Elzen, F. Geels and K. Green (eds) *System Innovation and the Transition to Sustainability*, London: Edward Elgar.

Hamilton, R. and Rollin Stott, J. (2004) 'Cycling: the risks', *Trauma*, 6: 161–168.

International Bicycle Fund (nd), 'Bicycle Statistics: Usage, Production, Sales, Import, Export,' available online at: www.ibike.org/library/statistics-data.htm (Accessed 07.07.12).

Jacobsen, P. (2003) 'Safety in numbers: more walkers and bicyclists, safer walking and bicycling', *Injury prevention*, 9 (3): 205.

Komanoff, C. (2004) 'Bicycling' in J. Cutler (ed.) *Encyclopedia of Energy*, New York: Elsevier.

Krizek, K., Handy, S. and Forsyth, A. (2009) 'Explaining changes in walking and bicycling behavior: challenges for transportation research', *Environment and Planning B*, 36 (4): 725–740.

London Travel Watch (2009) *Cycling in London*, London: London Travel Watch.

Lovelace, R., Beck, S., Watson, M. and Wild, S. (2011) 'Assessing the energy implications of replacing car trips with bicycle trips in Sheffield, UK', *Energy Policy*, 39 (4): 2075–2087.

MacKay, D. (2009) *Sustainable Energy – Without the hot air*, Cambridge: UIT.

Michaelowa, A. and Dransfeld, B. (2008) 'Greenhouse gas benefits of fighting obesity', *Ecological Economics*, 66 (2–3): 298–308.

Office of National Statistics (2004) *Greenhouse gas emissions from transport*, London: Office of National Statistics.

Pooley, C. and Turnbull, J. (2000) 'Modal choice and modal change: the journey to work in Britain since 1890', *Journal of Transport Geography*, 8: 11–24.

Pucher, J. and Buehler, R. (2007) 'At the frontiers of cycling: policy innovations in the Netherlands, Denmark, and Germany', *World Transport Policy and Practice*, 13 (3): 8–57.

Pucher, J. and Buehler, R. (2008) 'Making cycling irresistible: lessons from the Netherlands, Denmark and Germany', *Transport Reviews*, 28 (4): 495–528.

Rip, A. and Kemp, R. (1998), 'Technological Change' in S. Rayner and E. Malone (eds), *Human Choice and Climate Change: Resources and Technology*, Columbus, Ohio: Battelle Press.

Schatzki, T. (1996) *Social Practices: A Wittgensteinian Approach to Human Activity and the Social*, Cambridge: Cambridge University Press.

Sheller, M. and Urry, J. (2000) 'The city and the car', *International Journal of Urban and Regional Research*, 24 (4): 737–757.

Shove, E. (1998) 'Consuming automobility', Project SenceSusTech Report, 1, available online at: www.tcd.ie/ERC/pastprojects/carsdownloads/Consuming%20Automobility.pdf (Accessed 7.7.12).

Shove, E. (2003) *Comfort, Cleanliness and Convenience: The Social Organisation of Normality*, Oxford: Berg.

Shove, E. (2012) 'The shadowy side of innovation: unmaking and sustainability', *Technology Analysis and Strategic Management*, 24 (4): 363–375.
Shove, E. and Pantzar, M. (2005) 'Consumers, producers and practices: understanding the invention and reinvention of Nordic Walking', *Journal of Consumer Culture*, 5 (1): 43–64.
Shove, E. and Pantzar, M. (2007) 'Recruitment and reproduction: the careers and carriers of digital photography and floorball', *Human Affairs*, (2): 154.
Shove, E., Pantzar, M. and Watson, M. (2012) *The Dynamics of Social Practice*, London: Sage.
Shove, E., Watson, M., Hand, M. and Ingram, J. (2007) *The Design of Everyday Life*, Oxford: Berg.
Spinney, J. (2007) 'Cycling the city: non-place and the sensory construction of meaning in a mobile practice', in D. Horton, P.Cox and P. Rosen (eds) *Cycling and Society*, Aldershot: Ashgate, pp. 25–45.
Transport for London (2007) *London Travel Report 2007*, London: Transport for London.
Urry, J. (2004) 'The "system" of automobility', *Theory, Culture and Society*, 21 (4–5): 25.
Watson, M. and Shove, E. (2008) 'Product, Competence, Project and Practice', *Journal of Consumer Culture*, 8 (1): 69–89.
Wilson, S. (1973) 'Bicycle Technology', *Scientific American*, 228: 81–91.
Woodcock, J., Banister, P., Edwards, P. and Roberts, I. (2007) 'Energy and health 3 – energy and transport', *The Lancet*, 370 (9592): 1078–1088.
Woodcock, J., Edwards, P., Tonne, C., Armstrong, B., Ashiru, O., Banister, D., Beevers, S., Chalabi, Z., Chowdhury Z. and Cohen, A., (2009) 'Public health benefits of strategies to reduce greenhouse-gas emissions: urban land transport', *The Lancet*, 374 (9705): 1930–1943.
WWF (2008) One Planet Mobility: World Wide Fund for Nature, available online at: http://assets.wwf.org.uk/downloads/opm_report_final.pdf (Accessed 7.7.2012).

9 The making of electric cycling

Julien McHardy

In 2010, I took part in a test of electric bicycles run by a small public charity.[1] I joined the test as a volunteer and participant observer interested in tracing the practice of electric cycling. During the test, I kept coming up against normative tensions about what does and does not count as electric cycling. I found myself implicated in the test as a normal test rider and got to feel that the normalising of electric cycling requires the normalisation of bodies and things. The test, then, is a promising site for a discussion of normalisation on two counts. First, compared with other practices, electric cycling is more obviously a relational achievement in that bodies, and things, are intimately intertwined. Second, testing is an explicit part of the processes involved in establishing what counts as normal practice. Consequently, in this chapter, I focus on tensions in the test that arose around the negotiation of normal electric cycling, normal electric bicycles and normal electric cyclists.

Normalising electric cycling

Normalising the test

Electric bicycles are fitted with sensors and a battery powered motor, which propels the bike in tandem with the human rider. This electric support allows riders with different levels of fitness to travel faster and further then they could on push- bikes and, in so doing, potentially reconfigures what it means to cycle. Cycle Power,[2] the organisation that runs the test, champions electric cycling as an environmentally benign way of getting around. Emerging from the alternative solar mobility scene that played out around national and international solar car racing, Cycle Power has promoted electric cycling for over a decade. The long hours, extensive connections and determination of Felix Erad (the founding director) and his small team, has resulted in Cycle Power becoming a very unusual centre of expertise. The test is just one of several means by which Cycle Power aims to open up and establish a place for electric bicycling on already crowded roads, in legislation and in peoples' lives.

The test process began in the early 1990s, when Felix conducted a market review of the three models of electric bicycles that were commercially available at the time. He found that the manufacturer's technical data were at best unreliable and, since then, Cycle Power has developed increasingly sophisticated methods for testing and comparing performance. The recent electric bicycle boom promises to open up new markets for bicycle, automotive, battery and electronic companies and today, Cycle Power's services are in great demand by yesterday's sceptics.[3] The organisation's expertise, and its testing service, now play an important part in the making of normal electric bicycles and this also impacts on cyclists. Cycle Power, now arguably 'enacted' as a centre of expertise, conforms to normative pressures regarding the specification of valid knowledge and the meaning of reliable and relevant evaluation. Stories of the 'wild old days' at Cycle Power (when everything had to be improvised) enliven the lunch breaks, along with discussion about the precision (and 'hiccups') of current custom-developed test procedures and hardware. Of vital importance is the fact that the test is designed to evaluate electric bicycles in practice rather than under laboratory conditions. However, efforts to account for the particularity of actual cycling have to be weighed up against demands for clear cut facts about electric bicycles. Huge efforts have been made during recent years to create a professional, transparent and scientific test that can accommodate new models and meet demands for unbiased evaluation. No longer the private passion of an enthusiast, the test has been repositioned and it now figures as a service to be marketed and sold. This history results in an unusual conjunction of voluntary labour (the testers) and the scientific and economic demands of the test procedure. Stories that begin with, 'do you remember when…' perhaps indicate that testing has become somewhat normalised alongside the increasing standardisation of electric bicycles. The test, then, is not outside the processes of normalisation that it helps to enact, but is actively implicated in it.

Normalising electric bicycles

Walking along the rows of electric bicycles that Cycle Power keeps stored in a former barn, one can see how they have developed from home-made contraptions to commercial products. Early examples, from the 1970s, look somewhat 'home-brewed'; it is obvious that they are made of different parts, provisionally cobbled together. By contrast, more recent models represent distinct products in their own right, their branded 'skin' concealing the fact that they are still largely assembled from generic parts. Since sales figures in the European market started to increase a few years ago, they have been hailed as the successor to the mountain bike, the rise of which coincided with the global restructuring and growth of a declining bicycle industry in the early 1980s (Rosen 1993). Electric bicycles have begun to make regular appearances on streets, and in automotive fairs and shops, policy documents and articles; they are increasingly being seen as commercially viable alternatives to electric cars, which have consistently failed to materialise (Shukla and Kumar 2008).[4] At the moment, there are only a few motor systems available, but that is changing fast. Some of my informants predict a 'gold rush' in the industry that promises, as companies proliferate and consolidate, to blur the boundaries of bicycle manufacturing (from fieldnotes, 28.3.2010). In the process of circulating as commodities, electric bicycles have gone from being locally improvised arrangements, to becoming products that are governed (increasingly) by shared standards. Initially inspired by visions of sustainable movement, the focus is now on the business potential of electric mobility. Dreams of radically different modes of urban transportation have given way to scenarios in which electric cycling is integrated with existing infrastructures and industries. Despite these stabilising trends and tendencies, electric bikes are not yet smoothly configured; neither pushbike nor motorbike, they continue to disturb regulatory, legal, economic and technological categories and the informal distribution of vehicles and people on streets. I argue that processes of testing and evaluation have been under increased pressure to 'normalise' as electric bicycles have moved from the marginal realm of enthusiasts' projects to more explicitly capitalist forms of production.[5] The philosopher, Bruno Latour, argues that we need to understand how processes of purification and hybridisation constitute each other. For a thing to appear self-evident, it has to be cut off from its 'delegations and senders' (Latour 1993: 137). For a thing to appear 'pure', it has to be embedded in relations that are rendered invisible, so that the thing, divorced from its constitutive relations, can become self-evident or normal. In terms of the test, I argue that the networks looping through electric bicycles (and the test process itself) grow denser and wider as particular forms and models become normalised. Here, I think of normalisation as a dual process by which particular entities are woven through and cut off from their constitutive relations so that they can stand out as self-evident forms. Forms that become normal do so at the cost of other equally possible configurations. Having claimed that the pull of the anticipated gold rush smoothes out the edges and streamlines what counts as a normal electric bicycle, I now turn to the ways in which I was constituted as a normal rider during the test.

Normalising the electric cyclist

The motor and I speed up the slope, for the first time negotiating our efforts. Keeping an eye on the speedometer, I start to get a feel for how the sensor translates my efforts into electric support. We push each other as we work our way along the test track. I turn into a forest – tarmac gives way to concrete slabs and then to a dirt track. Having completed the first test ride, I turn back into the yard of the test organisation and enjoy the sensation when the gravel slides under the braked wheels. I press a button on the black box that, strapped to the handle bars and sensors, recorded our ride. After waiting for a bleep, as instructed, I pull out the USB stick and hand it to one of the two test managers. Like all other test riders, I will repeat this procedure after every ride for the following two weeks. GPS positions, numbers and graphs are brought up on the screen to quickly check for irregularities in the data patterns, which might indicate a failure of either measurement technology or rider. The prompt feedback allows the test

The making of electric cycling 135

managers to 'discipline' the technology or riders when the results transgress the margins of normality within which comparison is possible. Inspecting the digital traces of the ride, the test manager says:

> Over 300 watt rider output is too high. That is not normal cycling! You are supposed to cycle at the speed you would maintain during a tour not in some crazy sprint. Of course your output will vary according to the kind of bike you use but it should be within certain limits otherwise it will distort our measurements. It is not good if there is too much variation within the data.

During the project, the test managers considered each ride as an achievement of different riders, batteries, motors and frame geometries. This is an achievement that differs over time, as riders and batteries run out of juice during the day, change fitness and perform differently depending on changing conditions such as rest, temperature and weather. At the same time, too much variation has to be avoided given that the goal is to produce clear, comparable facts about electric bicycles. Staying with Latour, one could say that the testers aimed to measure every ride as a dynamic process of hybridisation while being committed to the production of self-evident facts that can circulate and that are independent of particular rides. In the end, riders have to be rendered similar because straight-forward results are more important in the test than attention to the particular qualities of different rides. Cycle Power's testing procedure relied on a cast of test riders, including an unemployed book restorer, a spiritual senior actor, an aspiring entrepreneur, a retired engineer, an ex-DJ, a local postwoman and me – a group as difficult to render similar as any. We were cared for, befriended and disciplined by a group of semi-full time staff and volunteers, including the two test managers who figure in these notes:

> On arrival we are weighed in the barn that houses the test bikes lined up along two walls of the room. On the opposite wall is a slightly leaking wood burning stove: batteries are sensitive to the cold and their performance varies considerably with temperature. The first thing I learn about my co-testers is their body mass. Moritz Genau, volunteer, cofounder and deputy director of Cycle Power who oversees the test together with Gerd Messer, the test manager, assures us that we have been chosen to resemble each other in terms of our physique, that we are similar enough not to disturb the stable field of comparison on which the electric bikes compete.

GENAU: 'We are comparing bikes not riders, that's why we have selected riders with a similar weight.'

Puzzled about our sudden similitude, we riders eye each other up. Genau perhaps picking up on our puzzlement continues:

GENAU: 'It would be good if you could adjust yourselves a little and ride in a similar fashion. Otherwise the data diverges too much and becomes difficult to interpret. It is good if the riders are not too different from each other.'

This, we learn, is a test of differences in models, not variances in temperature or in the rider's constitution. My task, as part of the fourteen-legged 'test human', is to represent diversity within the margins of a 'normal' cycling practice that has already been defined and inscribed in the test procedures and technologies.[6] Normal electric cycling is not the product of the test, but its precondition. In making this claim, I argue that the test managers required a somewhat reliable and thus necessarily pre-conceived notion of normal practice (and of normal practitioners) in order to compare the performance of different electric bicycles. And yet, the rider's status – are riders the passive point of reference in relation to which the bicycle's performance is measured or are they active elements in the making of electric cycling? – remained ambivalent throughout the test. Perhaps there is something about the distinctively dynamic constitution of electric cycling that makes it hard to pin down.

Electric cycling as a relational achievement

Over the course of the test, each test rider cycled thirty-five electric bicycles along three circular routes, in addition to three reference rides per track, test runs repeated due to technical or human 'failure' and those staged for photo and TV cameras only. During the test, the movement and efforts of 245 different human-machine-rider-weather-tarmac-breakfast-etc ... hybrids were measured and recorded by an array of machines that tracked speed, voltage, cadence and pedal power, alongside positions recorded by GPS.

> MESSER: 'Well, but you really do cycle different bikes differently and we are also looking to project these differences in the test.'
> Genau replies that while different bikes configure the rider differently that still does not mean that you have to sprint on one bike while you only cruise on another.

> I SAY: 'Well you have to tell me, I can also go slower, I don't mind, you just have to let me know. If it's best for the test I could go slower but this is how I would normally cycle.'
> 'That's not normal,' Genau insists. 'There is not only slow or fast, just cycle normally' she continues. 'We are looking for outputs between 100 and 200 watts, so it's not that we prescribe a certain value. It's actually quite a wide range.'

Messer and Genau cannot come to an agreement. Messer is keen to explore the varied riding practices that different bikes produce, whilst Genau insists on a margin of normality that should not be transgressed given that the aim is to compare the performance of different riders and bicycles. Genau also acknowledges that bicycles are constituted in relation to other entities: 'If bikes go that fast they come into a totally different power range and hence show a totally different "drivability" and are no longer comparable.'

An electric bicycle that goes at 30 km/hour is a different object compared to one that goes at 20 km/hour, and speed is partly dependent on the rider. Too fast, and the human-electric bike breaks away from the field of comparison to a different plane where it can no longer register in the test.

GENAU: I did take the STUD (one of the faster mountain bikes in the test) out for a ride yesterday to have a comparison and I went as fast as you would realistically ride such a bike – not on an extreme sprint where you pack a wallop but over a longer distance. In my case this was 180 [watts] and it is simply the case that the average pedelec rider rides between 150 and 200. If you buy a pedelec you tend to pedal along unhurriedly really.... If one buys oneself a pedelec one is hardly looking to race around on it.'

MESSER: 'I disagree. If you buy a STUD you also want to career around.'

GENAU: 'But Messer, here we have to take into consideration the normal pedelec rider and amongst those 150,000 bikes sold in Germany they are simply the majority.'

'Well...'

Later, the same day I return with one of the cruisers and Genau remarks:

'It all looks fine. 139 watts ... see, you can pedal slowly. There you go.'

MESSER: 'that's just the characteristic of the bike. You can only cruise it.'

Instances in which riders are 'disciplined' to perform within the thresholds of normal cycling recur during the test process. Some riders repeatedly exceed the threshold, while others consistently perform below the 100 watt boundary. This gives rise to a constant debate between Messer and Genau and marks some people out as less than perfect test riders, unable to pedal fast or slow enough. The conflict between Genau and Messer about how much variance can be allowed in the riders' performances is a conflict about what is measured and tested. Is it indeed a test of bikes, or of their dynamic and ongoing configuration in practice? In electric cycling, perhaps more obviously than in other practices, humans and non-humans are tightly intertwined. It is perhaps this intimacy of body-machine relations that makes it hard for the testers to hold human and non-human test subjects apart while attending to the practice of electric cycling, where the boundaries between both are blurred.[7] Elizabeth Shove argues that competence shifts between humans and machines as new tools and materials come into use (Shove *et al.* 2007), suggesting that such shifts occur over time as new practices or things enter into circulation. I would add that shifts in competence are the effect of ongoing compensatory movements and are dynamically and continuously negotiated. To some extent, such a strong focus on the dynamic ongoing relation between things and people challenges efforts to analyse changing practices by tracing the circulation of elements somewhat independently of their enactment (Pantzar and Shove 2010).[8] The intimate relations between riders and machines certainly challenged the testers' attempts to hold elements of electric cycling stable. In the next section I ask how electric cycling is stabilised in tension with conflicting accounts of its relational constitution.

Holding the human stable

Electric bicycles are difficult to test without test riders because their performance depends on the human element. However, the Cycle Power test is a test of electric bicycles and not of how different people might ride them. In order to come up with measurements that can be clearly attributed to these machines, the necessary human rider has to be stabilised so that he or she can be subtracted from the equation. The test aims to test electric cycling 'in the wild', but the demand for quantifiable and clearly attributed measurements requires that the test track, with its field and village roads, is configured as a laboratory of sorts. As a student of science and technology studies, I learned that all experimental systems leak and that even in strictly controlled laboratories, it takes an enormous amount of work to keep them afloat (see for example Rheinberger 1997). The testers do not need me to inform them that the shared dinners, drunken nights, fickle test technology, capricious and unpaid test riders and ever changing weather does not add up to a stable background against which electric bicycles can be measured. Their task is to find ways of holding some things stable *enough* in this shifting assemblage, so that measurements can be produced and circulated beyond the site of the test. The observation that the practical work involved in making such experimental fields is routinely purified, displaced and obscured in order to produce facts that can circulate beyond their site of production, has become one of the analytical staples of science and technology studies (for a classic version see Callon 1986). What is interesting here is how we, the test riders, participate in our own disappearance; how we make an effort to merge into the background and allow the bicycles to stand out as if independent both of our labour, and of assumptions about 'proper' electric cycling that we help to enact:

> On arrival at the test site we learned that the first route, the one leading up the village road and on to field tracks before it returns to the yard that now houses Cycle Power's facilities, simulates touring and should thus be cycled at normal speed. The second tour, leading to the neighbouring village of E., is comprised of several sections that simulate 'critical use situations: acceleration, continuous ascent, city traffic, start on a slope and motor-less ride.' The stretches between these sections are not recorded and, as far as the test is concerned, do not exist. In effect, the landscape has been indexed and cut up to represent moments in normal cycling. The virtual space that overlays the landscape still has to be realized through the circulation of surveyed human-electric-riders: sweat and voltage still have to be translated into data and plotted layer by layer, compromise by compromise.

On the first day we mark out the routes and the start of each section with plastic wrapped A4 signs: beacons attached to trees and light poles. Seven stop signs, passed twice on each trip simulate inner-city stop and go traffic in E. Everyone in the small village seems to be busy chopping wood. Wood is piled against walls, heaped up in yards and loaded on trailers. Apart from wood related dealings, E. is a quiet place stretched along three deserted roads. With the busy traffic of (future) mega cities in mind I stop and go on my way through the village. A stop sign is easily missed but clearly seen as a mistake on the resulting graph that, if correctly executed, forms fourteen indicative peaks. The people of E. watch us puzzled because to them the virtual plane on which the empty streets of their village are busy with urban traffic remains invisible. The man behind the garden fence is quite right to ask if I am racing against a ghost, for he and I occupy different roads.

John Law's discussion of research method is useful in conceptualising the relation between test and 'reality'. In 'Seeing Like a Survey', John Law polarises two understandings of method in science and social science. 'On the one hand it is usual to say that methods are techniques for describing reality. Alternatively, it is possible to say that they are practices that do not simply describe realities but also tend to enact these into being' (Law 2009: 239). Methods, he suggests, cannot capture the world as it is but interfere in ways that enact some realities and obscure others. The people of E. do not cycle as if they are in a survey and the reality we enact is obscure to them:

> ...like shadow players who cannot see the projection of their movements we have to anticipate the virtual traces of our rides. Our movements follow the demands of an invisible space that can be cut up, compared, evaluated in relation to or considered representative of normal touring, normal city traffic or normal ascents and descents. Eager to prove ourselves as capable participants – we are configured as good test subjects able to reproduce the circular graph that plots our labour. To cycle like a test rider you need to follow the routes that are marked but not fully described by the A4 signs. The test requires capable riders who know how to see city traffic on empty village roads.

As test riders we are disciplined by the technologies of the test and bound up in the reality and assumptions that it helps to enact. To cycle 'like a survey', that is to enact consistency, even partially, requires active negotiation by the test riders involved. We riders are not passive objects of discipline but actively bound up in the enactment of normal test subjects,[9] routes and rides. Our labour is deleted from test results, but we are part of and implicated in the vanishing act through which we disappear.

Turning into the coniferous forest I gather speed as I pass by the clearing that has become dear to me. Late afternoon light hangs in diffuse spears between regular trunks; on every ride I look forward to this little stretch and by now my body knows which path to follow, where to keep left and where to pace myself. Every time I cross this stretch of the circular route which has structured and used up the past days, the light and my cycling differ alike. Ideally, and as Genau has pointed out, I would cycle consistently regardless of the weather, the wind, the mood, the time of day, the company, the light, the increasing fitness or exhaustion and possibly [although that is contested, as we have seen] the kind of bike I ride.

Electric cycling, I have argued, is a radically relational achievement. How then are stability and comparability achieved if you can never cycle through the same field twice?

100 to 200 watts power output define normal cycling but how do you pedal with approximately 150 watts? We learn to ride in memory of recent and in anticipation of future evaluations, to pace our bodies and keep their performance within the margins of normal riding. Going up and down hill on different bikes and after consuming different meals – doing 150 watts is never quite the same achievement. Cycling with 150 watts does not involve steady power output but is instead related to the fluctuations of a changing landscape and the electric support this calls for. My body has become sensitised to the particular entanglement that is electric cycling, feeling the (sometimes) almost inaudible buzzing of the motor, constantly adjusting its effort.

I suggest that doing 150 watts in steadily changing configurations requires the dynamic distribution of competencies and labour. The stabilisation of 150 watts is consequently negotiated through the dynamic negotiation of subject/object divisions along a constantly moving edge. Where everything differs then, stability is not an effect of constancy but of dynamic negotiation: it is a distributed achievement. The work of stabilising a table on shifting ground depends on constant balancing acts rather than fixing; it requires improvisation rather than repair, and negotiation rather than maintenance. Stability does not depend on the impossible fixing of relational subject/object positions, but on their continual mutual adjustment. The considerable stability of a cyclist is an effect of motion. Drawing on this image, I suggest that stability is an achievement which is maintained through compensatory movements.[10]

As presented here, electric cycling is a relational achievement, taking shape within and between sites in which multiple versions of 'normality' circulate in tension.[11] Starting from the proposition that there is no one practice, I now ask how the versions of electric cycling that the test stabilised travel beyond the test site.

Re-relating the human

We test riders formed part of the compensation machine that gimballed the table on which electric bicycles could be measured and compared. Here, I briefly trace the route of the test data to show how facts about electric cycling travel beyond the immediate location of the test. The data from each test ride is entered into custom-written software. But how do numeric values such as battery voltage, left and right pedal pressure, torque and speed translate into judgements about what does and does not count as good technique? By the end of the test, the test managers can draw on several different data sets for each bike and each rider. These are aggregated according to procedures designed to produce precise averages for each bike, and to represent its performance with reference to twenty-seven different values, ranging from 'stability when parked', 'average city, tour and hill speed, range and support factor', to 'battery price'. At the end of this chain of translation (Callon 1986), the relational negotiations that constituted each separate ride are consolidated into facts about bicycle performance. At this point, the human rider has all but disappeared and we are left with a bunch of numbers. By themselves, these numbers say little about good forms of electric cycling, or indeed about anything very much at all. In order to make sense of the test data, the results (after abstraction from the practical relations in which they are produced), have to be related back to practice. This might be a test of electric bicycles, but in the end it addresses people who want to know what it is like to cycle this or that machine. Alone, cut off from these relations of production and appropriation, the figures mean nothing. The test report does further work in situating the results and also shows how testers make sense of their data:

> The 22 values measured in the Cycle Power test were then matched to the 16 [customer] wishes, thereby uniting (or merging) subjective values with objective measurements. Cycle Power established the list of sixteen customer demands based on years of experience, surveys and visionary thinking but in the end these categories are bound to remain somewhat subjective.

In a further step, the customer wishes are related to different electric bike product groups such as family, business, rehabilitation, wellness, city, classic or comfort. Each product group represents a different kind of user and is defined by a different weighting of the customer's wishes. Following a method adopted from the automotive industry, the performance of every bicycle is measured against the differently distributed wish-weighting of each product group. In the end, the winner in each category is the bicycle that measures up best against the particular wish profile of that group. In order to make sense, the measured values, (abstracted from practice) have to be re-related to imaginary practices. The test riders have, at this point, all but disappeared and cannot help. To make sense of the data, a second figuration of the human is introduced, namely the 'user'/customer.

In the rest of this section I argue that test riders and users are equally important figures in the development of normal electric cycling. Different imagined users and assumed wishes, whether those of families or businessmen, are required to make the test results intelligible. To summarise, test measurements are first cut off from the test riders' performances and then threaded back into practice through the figure of the user. Where test riders work to deliver abstract values, users function as a means of linking such values with imagined specificity.

Conclusion

In my discussion of the Cycle Power test, I have argued that practices such as electric cycling gather traction across multiple and potentially contradictory sites of enactment, of which the test is but one. The test can be understood as a crossing point of 'real' electric bicycles; assumptions about the 'real' world (in the form of future user groups that inform the test design, and the construction of particular realities in relation to which electric bicycles are evaluated), and the test reports which, once published, influence manufacturers and cyclists beyond the test. The bulk of this chapter has described how test riders, myself included, were implicated in the enactment of normal electric cycling through the test process. In looking at how forms of electric cycling travel beyond the test site I argued that the labour of the test riders can only become intelligible when it is looped through the figure of the user.[12] It is through a loop of cutting-from-actual and tying-into-imagined practices, that the test normalises particular forms of electric cycling. The test riders, once their labour is translated, are (metaphorically speaking) reduced to 'meat' – that is, to bodily material that can be cut up, measured, moved, processed and reconstituted. Users, on the other hand, are nothing but imagined future desires, until they are woven into the comparative matrix of test results. Because future users are imagined, and in a sense disembodied, they have to be re-embodied, situated and grounded in the thick of practical entanglements. The test process 'crafts' the bodies of future users by aggregating and managing data abstracted from the 'processed meat' that the test riders have become. The test-riders-reduced-to-numbers in turn only become intelligible when users link the test results to imaginary habitats beyond the test site. The abstract values to which the test riders are reduced (and the particular imaginaries of this or that user group) remain meaningless if they cannot be related. Both figures, test riders and users, gain credibility through the manner in which they are looped together. When they interlock, they form a machine that purifies and hybridises bodies in ways that enact particular forms of electric cycling, electric bicycles and electric cyclists as normal. Assumptions about what makes good electric cycling and hence what makes good electric cyclists (as well as assumptions about cities, roads, and traffic) are folded into the procedures through which test riders are disciplined and through which users are imagined.

Afterthought

In writing this chapter, I came up against a problem similar to that faced by the testers: namely, how to hold on to the relationality of practices while thinking about *the*, or *a*, practice as a relatively well-bounded entity. This might be an expression of my own limitations, but it may also point to a tension that others, who consider practice as a unit of analysis, might find relevant. I suggest that any attempt to conceptualise practice as an entity is in danger of skipping over the tensions that necessarily arise between normalising practices and the multiplicity of specific practical enactments. The techno-feminist thinker, Donna Haraway (amongst others) argues that close attention to situated relationality provides a means of foregrounding normative tensions and contradictions that might slip from view when practices are treated as ready-made packages (see, for example, Haraway 1988, 2008). Attention to relationality and normative tensions perhaps matters in testing electric bicycles and in theorising practices alike. After all, the ongoing delineation of normal practice, and of object and subject, is one way in which questions of power and politics play out in discussions of practice.

Acknowledgements

I would like to thank Gerd, Moritz and Felix and and all those at Cycle Power for their very generous support during my fieldwork. I would also like to thank Elizabeth Shove and Nicola Spurling for their patient and careful editing and Brigit Morris Colton, Natalie Gill and Philippa Olive for their comments on previous versions of this text.

Notes

1 The fieldwork I draw on in this chapter was conducted during 2010 in Germany as part of a wider PhD project on the making of electric bicycles as distributed things.
2 For reasons of anonymity, place, personal and company names have been replaced with pseudonyms.
3 This chapter focuses on the normalisation of electric cycling within the test, but it is important to recognise that the most common forms of this activity are precluded from the test from the outset. In 2011, 93 per cent of the almost thirty-one million electric bicycles sold worldwide were for Chinese consumption (Jamerson and Benjamin 2011: 34). The test is implicated in discourses of electric cycling that enact its division in the global north (namely Europe, Japan and the US) and China. Although vital for the future of the sector, Chinese forms are not captured or represented in the versions of electric cycling that are enacted in the test.
4 Considerable uptake has occurred in the Netherlands, Germany and Japan, and within the industry there is talk of a recent electric bicycle boom. However it is only in China that these bikes have become a ubiquitous mode of transport.
5 I do not suggest that voluntary labour, coupled with enthusiasm, exists outside capitalism. Rather, I point to a qualitative shift in which capital takes an interest in concerns and possibilities that were formerly dominated by other rationales than those of profitability.
6 See Madeleine Akrich for the argument that assumptions are inscribed into the design of things (Akrich 1992).
7 See Wallenborn's work in Chapter 10 of this volume for a more detailed discussion of how bodies are dynamically extended.
8 It is, of course, possible to conceptualise practices as both relational achievements and

entities as Shove suggests in her appropriation of Theodore Schatzki's distinction between practices as entities and practices of performances. Here, I only want to suggest that such a distinction has to be held in tension, because otherwise the normative dimension that is implicit in the delineation of practices slips from view.
9   I draw on Foucault's relational conception of power. My point about the riders' active participation in their own disappearance reiterates Foucault's argument that the subject is not simply subjected to power, but is one of power's prime instruments and most pertinent effects (see, for example, Foucault 1980: 98).
10 Latour argues that the double work of hidden assembly and official purification is a key characteristic of modern epistemology (1993). I share the view that the hidden work of relating should be made explicit so that it can become part of the political sphere. Feminists also argue that it is important to attend to the ways in which certain practices are rendered invisible so that others can stand out (see, for example, Mol and Mesman 1996). Anna Tsing argues that tensions do not cease when stability is established and that relations gain traction through friction (Tsing 2005).
11 Here I only hint at what Annemarie Mol calls multiplicity (Mol 2002).
12 In this chapter I focus on the ways in which normal practice is enacted in the test but it should be clear that assumptions about normal practices, lives and values are inscribed in users too. This argument is developed in detail by Nelly Oudshoorn and Trevor Pinch (Oudshoorn and Pinch 2003). The ways in which user groups are framed, for example, is bound up with the framing of cycling in terms of sustainability and environmental citizenship, the changing positioning of electric bicycles (for example, as auxiliary vehicles aimed at the elderly and now as fashionable, high status modes of mobility), as well as conceptions of health, wellbeing and aggregated data about disposable income, interests and attitudes.

References

Akrich, M. (1992) 'The De-Scription of Technical Objects' in W. Bijker and J. Law (eds) *Shaping Technology/Building Society: Studies in Sociotechnical Change*, Cambridge MA: MIT Press.

Callon, M. (1986) 'Some elements of a sociology of translation: domestication of the scallops and the fishermen of Saint Brieuc Bay' in J. Law (ed.) *Power, Action and Belief: A new Sociology of Knowledge?*, London: Routledge and Kegan Paul.

Foucault, M. (1980) *Power/Knowledge*, New York: Pantheon Books.

Jamerson, F. and Benjamin, E. (ed) (2011), *Electric Bikes: Worldwide Reports*. Naples, Florida, USA: Electric Battery Bicycle Company, available online at: www.ebwr.com (Accessed on 6.7.12).

Haraway, D. (1988) 'Situated knowledges: the science question in feminism and the privilege of partial perspective', *Feminist Studies*, 14 (3): 575–599.

Haraway, D. (2008) *When Species Meet*, Minneapolis: University of Minnesota Press.

Latour, B. (1993) *We Have Never Been Modern*, Cambridge MA: Harvard University Press.

Law, J. (2009) 'Seeing Like a Survey', *Cultural Sociology*, 3 (2): 239–256.

Mol, A. (2002) *The Body Multiple: Ontology in Medical Practice*, Durham: Duke University Press.

Mol, A. and Mesman, J. (1996) 'Neonatal food and the politics of theory: some questions of method', *Social Studies of Science*, 26 (2): 419–444.

Oudshoorn, N. and Pinch, T. (2003) *How Users Matter: The Co-construction of Users and Technologies*, Cambridge, MA: MIT Press.

Pantzar, M. and Shove, E. (2010) 'Understanding innovation in practice: a discussion of the production and re-production of Nordic Walking', *Technology Analysis and Strategic Management*, 22 (4): 447–461.

Rheinberger, H. (1997) *Toward a History of Epistemic Things: Synthesizing Proteins in the Test Tube*, Stanford: Stanford University Press.

Rosen, P. (1993) 'The Social Construction of Mountain Bikes', *Social Studies of Science*, 23: 479–513.

Shove, E., Watson, M., Hand, M. and Ingram, I. (2007) *The Design of Everyday Life*, Oxford: Berg Publishers.

Shukla, A. and Kumar, T. (2008) 'Materials for next-generation lithium batteries', *Current Science*, 94: 1–18.

Tsing, A. (2005) *Friction: An Ethnography of Global Connection*, Princeton, NJ: Princeton University Press.

10 Extended bodies and the geometry of practices

Grégoire Wallenborn

The unsustainable extension of our bodies

Late at night, after a 'glass of hot whisky and water', as he wrote, Samuel Butler sometimes ventured to speculate about the co-evolution of humanity and technology, the very process of humanisation. In the process of evolution, the human:

> learnt how he could, of his own forethought, add extra-corporaneous limbs to the members of his body and become not only a vertebrate mammal, but a vertebrate machinate mammal into the bargain. [...] The mind grew because the body grew – more things were perceived – more things were handled, and being handled became familiar. But this came about chiefly because there was a hand to handle with; without the hand there would be no handling; and no method of holding and examining is comparable to the human hand. [...] Were it not for this constant change in our physical powers, which our mechanical limbs have brought about, man would have long since apparently attained his limit of possibility; he would be a creature of as much fixity as the ants and bees. [...Machines] are to be regarded as the mode of development by which human organism is most especially advancing, and every fresh invention is to be considered as an additional member of the resources of the human body.
>
> (Butler 1912: online)

Butler goes further than Benjamin Franklin, who defined man as a 'toolmaking animal'. Being interested in the theory of evolution and in the development of machines in his century, Butler reflects on the organic growth of appliances and formulates a more radical proposition: becoming human depends on the creation of new unities of limbs and organs. Since this creation seems to be additive and endless, the search for sustainability is, in effect, a search for means of limiting these extensions.

Butler states that a human is defined by the possession of a series of mechanical limbs and organs (clothes, javelin, umbrella, watch, knife, pencil, book, spectacles, etc.), and by the forms of organisation that can be achieved with these prostheses. Butler thinks that each generation inherits prostheses preserved by natural selection, although he does not explain how selection happens. From the perspective of material culture, humanity evolves through the creation and reproduction of objects, tools and machines. Evolution is here taken in its broadest sense to refer to a continuous process of transformation. If it is possible to identify initial and final states (even arbitrarily), the process of change can be analysed in static terms: what has been transformed and what has remained constant? However, the real interest is in understanding the dynamic processes through which this transformation occurs.

Basalla (1988) explains the diversity of artefacts around us today in terms of technological evolution, arguing that fire, flint, animals, wheels, printing, steam and electricity have been added progressively to human practices. The hammer and the pin are positioned alongside the Internet and the nuclear power station. In this narrative, substitution and addition both occur such that technological evolution is viewed as a cumulative process, happening in an 'environment' populated by selecting agents that can be described in different terms: sociocultural, economic, military, psychological, etc. Patterns of technological evolution do not follow the well-ordered tree of the evolution of living beings.[1] Instead, different branches of technology are hybridised and yield new objects, for instance, as achieved through the conjunction of fire and water in the steam engine (Gras 2007).

The evolution of humanity is inseparable from the development of technology in that the extension of the body is at the core of what constitutes human beings as human. Material artefacts are integrated into human practices as soon as this process starts, meaning that our bodies have been extended to include new organs for billions of years. To reiterate, human practices always involve a body, more or less extended. Accordingly, humanity and materiality cannot be detached. However, bodily extension has become never-ending through various phases of industrial development. This is not a seamless process. The speciation of humanity, Butler continues, takes different forms amongst the rich and the poor. The rich person:

> alone possesses the full complement of limbs who stands at the summit of opulence, and we may assert with strictly scientific accuracy that the Rothschilds are the most astonishing organisms that the world has ever yet seen. For to the nerves or tissues, or whatever it be that answers to the helm of a rich man's desires, there is a whole army of limbs seen and unseen attachable: he may be reckoned by his horse-power – by the number of footpounds which he has money enough to set in motion.
>
> (Butler 1912: online)

Butler asserts that the difference between people lies in the organisation and extension of their limbs. Since the potential for extension is seemingly limitless – even the sky is no longer the limit, as the very rich can now afford space travel – the problem of sustainability is one of mastering this will or drive to extend.

It is well known that humanity has a problem of development. It is less common to view this as a problem of how extended bodies have evolved. For instance, if we compare cars and bicycles, we can assert that the former exhibits a kind of exoskeleton, whereas the latter has something like an endoskeleton. This is evident in how human bodies interact with objects: the object either envelops the human body (like a car) or is taken as an inner frame (like a bicycle). In both cases, the object can be regarded as an extension of the human body. The object is displaced and animated with, and by, a human body equipped with skills acquired through a more or less extensive period of learning. The coordination of movements and the organisation of perception consequently depend on both body and object, as a single assemblage. In the enactment of practices, perceptions are extended in different ways depending on the character of this assemblage. This is something that any cyclist knows and that the slightest tap on a moving car would show.

While the history of evolution shows the advantage of endoskeletons (vertebrates) over exoskeletons (invertebrates), humans have followed the reverse path during the last century. As car drivers, they shoulder hefty endoskeletons, 20 to 50 times heavier than the on-board human body. This is an admittedly extreme example, but one that illustrates the significance of bodily extension for sustainability. If individuals strive to 'possess the full complement' of their bodies, as Butler puts it, consumption patterns are clearly unsustainable: long before every body has achieved his or her full complement (which is always historically determined), the necessary stock of natural resources would have been depleted. The sustainability issue can be summed up as the following: flows of matter must decrease, some much more so than others. We know we have to be thrifty in our use of resources, and we know that the 'full' extension of some 20 per cent of contemporary human bodies already exceeds these limits. When extended bodies are too large or too resource-intensive, the 'footprint' or 'body print' of the practices they enact becomes unsustainable. If the extension of bodies is an unavoidable process, human history will come to an end. This is not the only possible path and, in analysing social practice in terms of extended bodies, one aim is to produce an account that is useful in helping to escape this fate. Such a description needs to show how new and different assemblages might be formed and how what we might think of as 'lighter' bodies could be constituted.

Extended bodies are very complex conjunctions: they include hard and soft organs; endo- and exoskeletons; specialised organs, skins and interfaces that together create layers of inner-ness and an outer-ness. The envelope of a body is its skin, and this is also true for an extended body. The skin of a body is more or less permeable, with holes enabling communication and different flows to come in and out. The same goes for a building in which the walls, as well as the inner atmosphere, create what is in effect an extended skin. Machines, infrastructure and networks (energies, water, communications, etc.) must also be included in the description of extended bodies, as well as the matters that flow through these channels. The use of others, of nature, of space and of time pertains to the constitution and definition of extended bodies. From this perspective it is not easy to see where the margins of an extended body lie. This, then, is a topic to which a description of extended bodies must attend.

In this chapter, my aim is to explore the link between practice and sustainability and to do so by describing practice geometries as the practice-based extension of our bodies. I suggest that understanding the transformation of these geometries could help in determining which extensions are problematic and which are not. In taking this approach I explore the idea that the unit of evolution is not the individual body, but practices that correspond to different extended bodies. As indicated above, a practice perspective, combined with a body-centred analysis, resolves the sustainability issue into an issue concerning the limits of bodily extension. Can we develop a theory of the evolution of extended bodies that is useful in identifying methods of redirecting or even transforming the trend towards yet more extensive extension?

In the next section I make a start by suggesting that the enactment of a practice is the performance of an extended body. Whereas practice theory tends to take the soft body of a person as the point of reference, treating people as the agents of integration, I start with a principle of symmetry that positions all bodies (human and non-human) on the same plane. I show that an analysis of extended bodies corresponds to an analysis of the material elements of practice, and confirm that practices are intrinsically social. Further ideas about the performance of the extended body are developed with reference to different concepts and authors. In the third section, I classify extended bodies in relation to five topographies: skins, fluxes, animals, plants and other human beings. These topographies describe the ways in which bodies are extended and then define the geometry of a practice as the singular intertwining of such topographies. In the fifth section, three problems are identified: where is the subject and where are matters of action and morality located in this analysis, and to what extent is extension 'automatic'? In section six, bodily extension is explained as an outcome of three possible processes: hybridisation, succession and delegation. I conclude with a discussion of how we might take better care of our extended bodies.

Practice as the performance of an extended body

Reckwitz defines practice as 'the regular, skilful "performance" of (human) bodies' (2002: 251). In this account, bodily activities enable the emergence of coordinated entities that include routinised mental and emotional activities, objects and knowledge. The practice as an entity ties these heterogeneous elements together. In order to explore the relations that constitute a practice, or an extended body, I will start from a principle of symmetry, describing connections between heterogenous elements before exploring the variability of the 'extended body' concept and considering the type of philosophy to which these ideas relate.

It is not quite right to equate a practice with an extended body in that extended bodies are inherently dynamic, always living and always changing. It is therefore more accurate to say that the enactment of a practice is the performance of an extended body. The performance of a practice can therefore be understood as an identifiable configuration in the flux of individuals' relations with each other (e.g. when two persons speak to each other). A practice always involves an 'equipped body', because such a configuration always includes materials beyond the naked body. Schatzki (1996) describes practices in terms of intention or teleo-affective structure, but materials do not figure in this part of his analysis. By contrast, my reference to the extended body highlights the material elements of practice. The heterogeneous elements of the extended body are actively linked together in a practice. In other words, an extended body is an assemblage that 'functions' on a more or less regular basis. What can we learn by identifying the performance of a practice with the functioning of a unique and singular extended body?

One feature of such an approach is that it establishes practices and extended bodies as inescapably social: they are so in that extended bodies encompass objects (themselves the condensation of multiple social relations) and other people. The body is social in the sense that it is an 'association of humans and non humans' (Latour 1999: 286). In addition, a practice (as entity) can never be individual, since it entails a large number of repetitions by a large number of people. From the extended body perspective, the performance of a practice – bodily activity – is conceived of as an emergent property: it is the visible face of a practice as entity. 'Bodily activity is the appearance of mind, and mind is the expressed bodily activity' (Schatzki 1996: 54). Behaviour, defined as observable action, emerges from the performance of a practice, that is, from the active linkage of heterogeneous elements. Behaviour is a popular concept, partly because of its visibility (or perceptibility). For example, changes in movement can be seen and detected. However, we need to go beyond these outward manifestations if we are to grasp and characterise the distribution of differences and similarities in practice. Looking at observable bodily activity is insufficient if we are to understand the evolution of extended bodies. Looking at an extended body is, in effect, a means of looking at how the different components of this body move and are articulated together with the settings and networks that are also part of it.

The material aspect is evident from the start in that an extended body consists of an integration of human bodies, objects and specific elements of the setting. The biker combines with the bike, and the road itself is part of the functioning of the extended body. The biker and the path can have variable geometries, since they have to be compatible. The racing bicycle needs a smooth road, whereas the mountain bike is adapted to muddy paths. As this example suggests, there is some kind of mutual selection between the different elements of the extended body. Furthermore, the extended body is not just a combination of solid elements as a mechanism might be. The place of the human body and the nature of its extension are rarely that clear.

The extension of the body depends on what is considered as belonging to the body, and different points of view are possible. While some think of humans as toolmakers, capable of creating new limbs and organs, others view the extended body in less mechanical terms: for example, writing about sociotechnical hybrids, cyborgs, desiring-machines and the organisation of fluxes. In such accounts, material evolution proceeds through the hybridisation of natural and cultural elements, new combinations of which emerge from a tension between innovation and habit. It is useful to review some of the ways in which extended bodies have been conceived.

The long introductory quotation from Butler resonates curiously well with Spinoza:

> no one has hitherto laid down the limits to the powers of the body, that is, no one has as yet been taught by experience what the body can accomplish solely by the laws of nature, in so far as she is regarded as extension.
> (Spinoza *Ethics III*, prop. 2: online)

Spinoza escapes the Cartesian mind/body dualism by arguing that body and mind are not two separated entities, but two modes (or points of view) of the same phenomenon. This implies that the mind is extended in parallel to the body and is distributed in the body, much in the way that intelligence is distributed between different coordinated agents.

Also designed to counter mind/body dualism, cyborgs and desiring- machines have been conceived of and introduced initially as a means of escaping the tendency to naturalise feminine/masculine features and anchor them in biological bodies (Haraway 1991). However, the cyborg is more than a feminist weapon; as a hybrid machine/organism it is a compound of human, animal and electronic elements:

> The machine is not an it to be animated, worshipped, and dominated. The machine is us, our processes, an aspect of our embodiment. We can be responsible for machines; they do not dominate or threaten us. We are responsible for boundaries; we are they.
> (Haraway 1991: 180)

The figure of the cyborg includes all possible extended bodies. As such, any extended body is an actualisation of a cyborg. A practice is, then, a transient cyborg, the limits and geometry of which are constantly changing as objects and organs are re-distributed in specific ways in relation to a given situation. The objects included in an extended body help specify the practice but do not determine it: for example, the same object – a knife or a computer – can sometimes be used to perform different practices.

In writing about 'desiring-machines', Deleuze and Guattari (1983) also go beyond the idea of the tool as a mechanical extension of the body. The deleuzo-guattarian machine is a bricolage that works with, and assembles, heterogeneous elements. The functioning of this desiring-machine is similar to the performance of an extended body, but in this case there is a clear emphasis on experimentation. The desiring-machine is a nexus of different parts, again including human, animal, plant, land, tool and machine. However, the result is not a pre-established 'structure': instead the extended body is continually organising itself, linking and de-linking multiple elements, but never achieving any lasting unity. The concept of desiring-machine emphasises this vital aspect of ongoing production. As it proceeds, the machine undoes and redoes itself; it stops, goes, and tinkers with its own extension. The desire is that which 'machines' the heterogeneous elements involved in this unfolding process. Accordingly, desire is a material process that produces reality. The desire is immanent to the machine, and desiring-machines constitute the fabric of social practices.

Linking the idea of an extended body to that of a desiring-machine underlines the experimental aspect of practice: bodily extensions are not made in order to supply a deficient organism with new limbs, but to create new situations and engender a new extended body (Raunig 2010). In being extended, the material parts enable the functioning of this new body. At the same time, it is this functioning that unites the multiple elements, forming a transient entity which is the locus of doing and meaning.

Bennett (2010) develops another way of seeing extended bodies, doing so with explicitly ecological and sustainable aims in mind. Bennett's opening proposition is that matter is not only solid; it is also fluid, energetic, vibrant and evanescent. Extended bodies are consequently constituted through flows including those of air, light, drinks, food, electricity, waves, energy, signs, etc. As such, they figure as agentic assemblages of microbes, animals, plants, metals, chemicals, word-sounds and the like. Bennett's 'heterogeneous monism of vibrant bodies' draws on diverse philosophical resources: Spinoza's 'natura naturans', Nietzsche's world of energetic flows, Deleuze and Guattari's vibratory cosmos, Bergson's creative evolution and Michel Serres' vortices and eddies. This account makes much of the fact that humans are made of non-human bodies and that the way matter is distributed through an extended body should not be taken for granted. A key insight is that networks (such as roads, electricity and water supply systems) are constitutive parts of extended bodies.

Winance's (2006) discussion of wheelchairs echoes many of the points made above. Again the claim is that the wheelchair actively fabricates a new person, forming a new assemblage of human and non-human entities. Winance uses the idea of the extended body to describe the abilities and disabilities of the new collective made of a paraplegic, a specific wheelchair and the disposition of things, including the hands of other humans. The further point is that in so far as wheelchairs enable and prevent movement in specific places (kitchen, car, lift, etc.), this equipment makes its own world. These relations are such that there are many different wheelchairs, and just as many corresponding worlds. Winance shows that this kind of world-making is the result of a long process of negotiation between a human body, emotional activities, objects and others. The process of appropriating a new prosthesis is thus one involving the reciprocal domestication and realignment of different parts of the extended body.

To sum up, the extended-body approach is consistent with a view of practice as a process in which heterogeneous elements are actively integrated and in which the distinction between object and subject is blurred. This approach draws inspiration from those schools of philosophy that challenge dualism in all its forms: mind/body, symbolic/material, culture/nature, human/non-human, individual/totality (Schatzki 1996). As the extended body is neither structure nor habitus (mental disposition), further understanding depends on concepts that are not substantialist, but that refer to relations and processes. Each philosophy is sustained by an ontology that specifies the types of 'being' taken into account. In practice theory, the beings – the extended bodies – under scrutiny should be thought of in dynamic terms: they are as changing as the practices themselves.

This line of thought is not without its problems. For example, where does intention lie, and does the human body have a special place within the extended form? Before tackling some of these issues, I want to show that extended bodies can be analytically separated and discussed with reference to five different 'topographies'. This analysis will help in describing the geometry of a practice and in better understanding what more sustainable extended bodies might look like.

Topographies of extended bodies

Any given practice is unique if it is described in detail. And so is an extended body: one is always slightly different from another. At the same time, extended bodies can, and should, be classified so that they can be identified, described and thought about. One method of classifying extended bodies would be to distinguish between the different functions of the material elements, for example treating each as if it were an organ contributing to the operation of the whole. However, this rather 'functional' view is of limited value in capturing the dynamic relation between the different parts and in describing the changing margins of the extended body. Since a good description should be able to characterise types of extension my aim is to classify extended bodies in terms of their configuration or geometry.

Here, geometry is understood in contrast to arithmetic. Statistics, an elaborated version of arithmetic, deals in countable objects and consequently treats individuals as bounded entities. By contrast, geometry is related to topology, morphology, relationships and intensities. Geometry emphasises features of spatial and material extension. The geometry of an extended body is what makes it recognisable (a mobile phone cannot be understood without the networks that feed it). From this point of view, practices have different shapes or geometries. Exercises, like those of calculating carbon footprints, depend on the Cartesian reduction of geometry to arithmetic, a necessary step for those interested in counting and comparing numbers.

Geometry can be used to describe the configuration of an extended body and its corresponding practice. It can also be used to discern and characterise the different *topographies* in a practice, namely the different types of *places* (topoi) in which a body is extended. Topographies are regular patterns that can be observed in social practices. Literally, topography delineates the outlines of practice/body extensions in different domains. These multiple topographies can be analytically superimposed and combined to produce a composite portrait of practice geometry. In reality, a practice geometry consists of a dynamic amalgamation of topographies that are inseparably intertwined. Even so, it is useful to distinguish between topographical types as means of identifying the different ways in which geometries extend. The following paragraphs describe five such types – skins, fluxes, animals, plants and other humans – and show how these combine in shaping practice geometries.

Skins mark and create the difference between an outside and an inside, this being a climate or ambience within which the extended body acts. From clothes to homes and from caves to skyscrapers, skins have continuously evolved. Skins provide protection from outside 'threats' and determine what can or cannot enter the extended body. Skins have holes of different sizes that allow different fluxes to come in and out. In this context, keys are special devices that distribute skins and bodies, and membranes are filtering boundaries.

The second topography, flux, is the complement of skin and boundary making. It is grounded in the Heraclitean view of nature as a flux (Nietschze) or as a viscous flow (Serres). In thermodynamic terms, the human body is an open system, a permanent restoration of atomic assemblages, a dissipative structure constituted by flows of matter, energy and signs as well as by food, water, air, light, excreta, words, etc. To this the extended body adds further networks of provision: electricity, water, telecom, roads, etc. In other words, the extended body has different dimensions of flux, along with associated milieux including electromagnetic fields, electrical networks, roads and service stations, sanitation ducts, and so on (Simondon 1980). The fact that water is heated to body temperature before showering shows that the shower water belongs to the extended body. The water that drains from the shower tray is, in turn, an outcome of the extended body. Just as the human body is constituted of material flows, so the extended body is formed of the different fluxes which feed it. The skin preserves the margins of the body (it keeps our insides in) but allows aspects of the extended body to flow through. The limits of an extended body are therefore fuzzy. In what sense can we say that an external flux belongs to an extended body? The incorporation of a flux is an event: it might not have happened, but once it has, the effect is to reconfigure the extended body. This suggests that we should think of the appropriation of fluxes as a process of incorporation by the extended body. In this sense, the road comes through the driven car considered as an extended body. Fluxes are realised – and real in their effects – when they connect with each other and combine in (re-)making extended bodies.

The third topography, termed the animal body, is a mobile entity: it has a relatively defined shape and can move without being dismantled. The animal body is usually extended through tools, instruments and machines. Many tools are, for instance, extensions of the hand. Instruments change what can be sensed and done and so change the 'scope' of a practice. A machine comprises a source of power, either internal or external. This can come from the human body, as with a bicycle, or from some external source like a domesticated animal or an engine. A rider with a suitably equipped horse forms a remarkable extended animal body.[2] Anything that can be considered 'on board equipment' is part of the animal body.

In contrast to animals, which can move, plants are living beings that have a fixed location (Hallé 2002). This implies a different logic of extension. Plants cannot travel, but they are interconnected through chemical signals, by their roots or by third parties like mushroom mycelium. Complementarity and symbiosis are features of this, the vegetal world. Whereas animals extend with the aid of prostheses, plants do so through grafting and hybridisation. With plants, the notion of individuality is distinctly unclear. Is a swathe of grass a unique being or a multitude of separate blades? Due to the rhizomatic connections involved, this is not an easy question to answer. Similarly, the descendants of a plant cutting have the same genetic material as their parent, but do they also have an identity of their own?[3] As these comments suggest, individuality is an animal notion, based on the potential for movement. The vegetal body extends through connections to other bodies. In this regard, information and communication technologies help us become more vegetal. Typical vegetal extensions of the human body work through the fingers, for example, through the operation of phones and computers. As in a nervous system, these local interactions connect with distant points, linked through multiple networks. A yell or a pheromone release, for example, is a burst of signs as long as there is a detector in place to receive and interpret. In a nutshell, the vegetal topography describes the web of interactions that tie an extended body into its associated milieu.

The fifth and last topography is probably the most difficult to describe in detail because it involves several human bodies. Up to now, I have considered practices that involve only one human body. However, when other humans are involved in the same practice, for example, in dancing, chatting, sporting, loving or playing, it is possible to have different accounts of the same extended-body performance. In such situations, much depends on the *point of view*. In football, as in any other collective game, there are several points of view. From a player's perspective, the whole game is a 'body' incorporating the ball and all the other players (including the referee) and defined by strict spatial limits and rules. If asked, each player would probably provide a different account of how his or her 'own' body was extended during the match. At the same time, this whole extended body, composed of multiple people running, becomes a single entity when unified by an external point of view, such as that of a spectator. The agency of a multiple-human-body is particularly complex since actions and signs from any one component human body are constantly interpreted by the other human bodies involved. Agents anticipate what the other(s) will do and say, and the capacity to adopt another's point of view complicates internal relationships within the extended body. Another issue with the multiple-human-body is its variable relation to other topographies. Multiple bodies can share a common skin, for instance if all inhabit the same building. Fluxes are also variably shared. In addition, asymmetric situations frequently arise: for example, a car driver and passenger have different animal and vegetal topographies, but share a common skin.

Practice geometry as the intertwining of different topographies

It is possible to make analytic distinctions between these five topographies but, in reality, they are intertwined in practice. The analysis sketched above is nonetheless useful in that it allows us to consider forms of addition and superimposition that occur within extended bodies. For instance, mobile appliances (cellular or computer wireless) are hybrids of animal and vegetal topographies. Embedded electronics combined with networks enable a connection between the rhizomatic and the animal body. The acts of phoning or using the Internet exemplify rhizomatic modes of existence but, when the equipment is carried 'on board', the extended body is hybridised. This hybridisation is recent since it requires an associated electromagnetic milieu into which ready-made detectors can 'plug' almost anywhere (this has been the case for some time, as with transistor radios) and into which emitters can connect (this needs a dense grid of antennae). Electronics opens up a wide range of new possibilities and could increase the integration of vegetal and animal parts, perhaps in the form of new cyborg configurations. These might involve new holes in the skins of extended bodies; new sensors might permit new ways of regulating and perhaps intensifying interactions with other extended bodies, and the combined result might be heavier or lighter animal bodies that require more or less resource use and that produce different fluxes of material and electromagnetic pollution.

In thinking about how topographies intersect it is important to think about the extent and definition of an extended body. Machines are a special case of an extended body, in that some can work without a human body providing there is an external source of power. Consider the example of a home. From the outside, the extended body is characterised by a skin (the walls) and different incoming and outgoing fluxes. But from the inside, the home is more like a machine of machines. The intensity of relations between these machines depends, to some extent, on the degree to which the skin of the home is permeable. At first sight, an inhabited off-grid canal boat is a good example of an animal extended body. It is a movable entity but one that contains a number of inner machines. The fact of being off grid means that these inner machines need to be carefully coordinated: there is a delicate, vegetal, symbiosis between the different parts of this extended body. Since resources are necessarily limited within this self-contained entity, special attention has to be paid to fluxes (to fluids and to power for instance, provided by solar panels or a small wind turbine). In situations like these, the different topographies are very densely interwoven.

The five topographies do not have a hierarchical order, thus the rational or the animal are not in some sense above the vegetative, as is the case in an Aristotelian view. Instead, each topography relates to a different characteristic of the extended body. The geometry of a practice is the assemblage of these different topographies, but there is no 'pure' geometry in the sense that an extended body might be described in terms of a single topography. Skin and fluxes are always involved, and can be more or less extended – these concepts relate to margins and flows. And animal and vegetal topographies characterise the active functioning of heterogenous parts and their relation to the environment – what I term a regime of action. For example, the animal body concept describes solid parts and the relation between them (in terms of distance, speed, etc.) By contrast, the vegetal body concept describes interaction with different milieu, linked through fluids and interpretations of signs. These animal and vegetal forms relate to skin and flux in characteristically different ways. Skins indicate the margins (the inside and the outside) of the animal body and are a necessary but not sufficient condition of its mobility. Although vegetal bodies have skins, these are often harder to detect. Whereas skins separate animal bodies from their environment, skins connect plant bodies to their milieu. The body is at first sight identified by its skin whereas fluxes show how the body evolves and is linked to its environment. Fluxes of matter are therefore more closely linked to vegetal bodies in as much as fluxes have a fluid feature. In this sense, flux and vegetal topographies are two sides of the same process: the fixed body captures, assimilates and expels fluxes of matter, energy and signs, whereas the rhizomes and branches carry and organise these fluxes within and between associated milieux.

The topography including other human beings implies a third regime of action in addition to the animal and vegetal forms described above. This regime is based on the interaction between entities that have their own bodies, skills and meanings, aspects of which are more or less shared within the extended body.

To sum up, I suggest that all practices could be considered and analysed as combinations of these five topographies. The example of cooking – reaching for ingredients, combining and heating them in a pan, using first a knife, then a spoon, etc. – shows that while the topographies do not vary, the geometry of the practice changes continuously as it is being performed. Taking a longer term view, it might be possible to represent the history of a practice (e.g. breakfast) as the evolution of the geometries or topographies involved.

Subjects, agency and habits

If these ideas are to be plausible, we need to explain how bodily extension occurs. For this purpose, I need to answer three questions: (1) If a subject experiences a succession of extended bodies, how can I say that it is still the same subject? (2) How is the agency of an extended body characterised? (3) To what degree is the extension automatic? As we shall see, these questions are entangled.

Extended bodies are transient assemblages that are also meaningful units of practice. It is in these transitory configurations of human bodies and objects that human life is experienced. The extended body is a living body. This living body has at least two relatively enduring characteristics: a human body and a memory. All human bodies decay but what of extended bodies, do they also age? We can here distinguish between the ontogenesis and the phylogenesis of the extended body. Ontogenesis implies that a human practice always involves a body, more or less extended and, conversely, that a human body is (almost) always equipped.

Phylogenesis represents human evolution as the creation of new unities of limbs and organs. At the human scale, phylogenesis is the development and specialisation of practices, whereas ontogenesis represents the more stable history of our innate corporal and cognitive capacities. Although the parts of extended bodies progressively deteriorate, the extended body as a practice does not. Practices are not worked by time in the same way as human bodies: they are reproduced through imitation and through the evolution of identifiable configurations.

If practices are the meaningful units of life, it makes sense to conceive of the subject as the passage of successive extended bodies. The identity of the subject is produced and reproduced in and through practices. But in this case, where are the persons morally responsible for their actions and able to relate to others? As morality is a modality of action, we have to turn to the notion of agency if we are to address this question.

Latour argues that agency is distributed between humans and non-humans (Latour 1999). How does this distribution happen in an extended body? From the perspective of an extended body, agency is immediately material; teleoaffective, understandable and rule governed. However, activity and thought are distributed along the articulations and fluxes of the extended body. The mind is, for instance, distributed in the extended body in that intelligence can be distributed between different coordinated agents. In the vegetal regime of action, agency lies in the connection of the body to its environment: actions and flows are channelled and perceptions stretch across the entire extended body. Nevertheless, perception depends on the point of view: some organs are more obvious than others. Since the practice is the actualisation of an extended body, the question of agency is transformed into one that has to do with the composition and coordination of that body. The agency of the extended body depends, in turn, on how relevant topographies are articulated, and on how objects are incorporated. The agency of a practice consequently lies in the articulation of the heterogeneous parts of the extended body.

Besides having a body, the human subject is also characterised by a memory and the ability to recount life events. Just as agency is distributed, so the memory of a practice is distributed among the different parts of the extended body. When an extended body is re-enacted, its memory is recovered. 'Sleeping organs' can be reactivated when memory re-connects these elements. To the extent that an extended body exists as a singular, one-off experience (a moment in a practice), vitality is distributed through matter and objects. The process of life consequently consists of ongoing experimentation with and through a stream of new extended bodies. Like agency, morality is distributed and delegated to the different components of the extended body (Latour 1999). But that does not override the point that the human subject has the capacity to partially guide such experiments, and to configure new extended body arrangements or repeat stabilised assemblages.

Understanding the formation of habit is important because many practices are performed in a regular, semi-automatic way. Practices emerge and replicate themselves in different spatiotemporal frames. Domestic and daily practices are, for example, characterised by a high degree of repetition involving the same body configurations. The persistence of relevant aspects of the extended body (and of its related geometries) is critical, but this is not enough to ensure that practices are recurrently propagated. Tacit knowledge is an essential component of the extended body, and can only be transmitted through imitation. As such, it depends on the coexistence of other human bodies (Collins 1974). A technology always involves a 'body technique', which is learned through the repetitive performance of that technology. To illustrate this point I recently experimented with a new extended body that challenged my previously embodied habits. This involved trying to steer a tricyle with my feet while pedalling with those same feet. After two hours, I was not yet sure how I should command my extended body, but by the next day my skills were much improved. This kind of learning process, or embodiment of a new practice, is well known. The body technique creates and stabilises relations between different kinds of objects and bodies, including one's body and the bodies of others. These relations are not only technical since they can be psycho-affective or ludic (playful). Rather than using the term 'body scheme', Warnier (1999) prefers to use 'driving conduct' (*conduite motrice*) to signify this active relation to materials.

> Body synthesis (or body scheme) is the synthetic and dynamic perception that a subject has of herself, of her driving conducts, and of her space-time position. It requires the senses in their relation to her own body and to the material culture. The synthesis results from a learning process that happens all life long. It shows a great deal of individual, cultural and social variability, while assuring the subject continuity in her relation to the environment. It successively dilates and retracts in order to integrate multiple objects (car, appliances, clothes, sport equipment, etc.) in the subject's driving conduct.
> (Warnier 1999: 27)

The behaviour of the extended body stands in tension, positioned between innovation and habit. A habit is thus an experience crystallised into an extended body.

A practice approach shows that the human subject does not aim to extend his or her body for its own sake, or to become a complete organism, but to perform known configurations (habit) or to test new situations (innovation, experimentation). Does this process have any direction? Following Spinoza, Bennett states that 'bodies strive to enhance their power of activity by forging alliances with other bodies in their vicinity' (Bennett 2004: 352). The goal, according to Spinoza's ethics, is to increase the degree of self-direction regarding one's encounters. Whether we go along with that conclusion or not, the story of the extension of bodies provides a set of ideas with which to reflect upon our encounters with matter.

Three means of using the extended body concept

To summarise, the concept of an *extended body* allows us to explore the links between practice and sustainability. Extended bodies give concrete shapes to practices and practices, in turn, cut across individuals. Understanding how this works is a matter of understanding processes, events and relations: it is not a matter of substances or essences. After this long detour through extended bodies and their topographies, I will now try to give some hints as to how such an account might help in moving towards sustainability. For sustainability, the issue is one of matter, and from this point of view an obvious response is to curtail aspects of material extension. However, that goes against a tendency observed since the start of the Industrial Revolution, this being towards maximum extension. To understand how this has come about I need to say a few words about how extended bodies change over time.

To the extent that the performance of a practice is the functioning of an *extended* body, aspects of spatial extension are critical in that they involve and allow the intertwining of different *topographies*. Material extension can result from three different processes: hybridisation, succession and delegation. We have already seen how hybridisation occurs through the combination of different topographies (skin, flux, animal, vegetal and other humans). Each topography has its own trajectory. In addition, and as already discussed, the superimposition of topographies is made possible through networks, systems of provision and associated milieux. For example, cars have developed as integrated extended bodies because infrastructures (especially cities) have been organised around them. In addition to these kinds of surface networks there are underground systems of ducts and pipes. Above ground, the air is now criss-crossed by electromagnetic waves. The creation of new milieux of this kind permits the superimposition and hybridisation of different topographies. Understanding hybridisation is consequently important for understanding how the mobility of humans and non-humans connects to infrastructures.

Besides the point that extension changes the potential for superimposing topographies, the multiplication of extended bodies is of direct significance for questions of sustainability. Each time a new object comes on to the market, either a practice changes or a new practice appears. When the effect is one of addition rather than substitution, the sheer number of extended bodies increases. The extent to which a practice is embedded depends on the number and distribution of practitioners and the frequency of their performances. The potential to pack more practices into a day relates to the fact that relevant infrastructures are managed and maintained by others (humans and non-humans). Parts of my extended bodies (house, car, machines, infrastructures, etc.) are kept in good condition by others so that I can behave like an autonomous animal body. Such arrangements mean that I encounter many different objects and situations within a given period – an hour, a day or a year. These multiple encounters are directly related to the increased number of objects that populate our lives. They also have to do with the potential for delegation. People are able to perform practices in parallel within an extended body because they are able to delegate some tasks to machines. The idea that objects somehow take the place of human labour (labour saving devices) does not make sense in terms of an analysis of the extended body. From this point of view, a functioning washing machine or oven are parts of the home, viewed as an extended body. The concatenation and simultaneity of practices in the same extended body, for example, the home, draws attention to the importance of coordination and the significance of reliable external sources of power.

Implications and conclusions

The contention that practices are enacted by extended bodies suggests that practices cannot be simply dematerialised: the material is integral to the extended body and its performance. More broadly, the tendency towards bodies that are extended further and towards a world in which there are simply more extended bodies (not necessarily related to an increase in population) helps explain why industrialised countries consume so many resources. How might these ideas be translated into suggestions for sustainable policy, beyond already familiar advice to live slower and more localised lives? In this last part, I show how the five topographies lead to new viewpoints about the way we care for things and living bodies.

First, there is a need to care for and to develop new skins. Consider skins that are related to comfort. Considerable resources are today required to condition space (in terms of temperature, humidity, light, quietness, etc.) within the envelope of a building skin but we know that comfort can be obtained in different ways, for example through clothing, itself an extended skin. Other questions relate to the distribution of skins. For instance, land planning might be thought of as a process of establishing urban and rural skins, and of managing connections and forms of circulation between them. On the other hand, when precious fluxes enter an empty skin, the result is obviously wasteful. Equally, existing skins might be 'larger' than we need, hence imagining future skins that are more adaptable, and that have a variable geometry.

Second, fluxes are rarely considered. Instead, dominant discourses focus on questions of access to stocks of material or energy. This metaphysics of stock leads to a view in which environments are expected to deliver relevant resources and products at the right place and at the right time. By contrast, the metaphysics of flux suggests that extended bodies can and should be adapted to actual fluxes. For instance, renewable energy is produced when it is windy or sunny. At moments when energy is produced, extended bodies might be activated, but scaled back when energy is scarce. Attending to fluxes is not only a matter of paying attention to the variable intensity of practices/extended bodies. It also reminds us to consider beneficial and harmful flows: what pleasure or damage is caused by the products that go through the gates of my skin in both directions?

Third, the development of animal parts, through the creation of new solidly bounded entities, has led to increasingly complex forms of coordination and delegation. The extension of animal topographies depends on resources such as space, time, energy, water and materials of different kinds. A straightforward recommendation is to reduce the animal parts of extended bodies. This is especially important since the animal regime of action depends on specific concepts of identity and ownership. To rethink the animal topography is to question the regime of private property: what objects should belong to one human body, and what could be shared among different human bodies? As animal topography and mobility are directly linked, taking care of our animal bodies depends on paying attention to their movement, speed and direction of travel.

Fourth, certain vegetal parts of extended bodies deserve deliberate cultivation. This suggestion should not be confused with a plea for vegetarianism or a 'rooted' existence. Rather, a vegetal orientation is, in essence, a matter of focusing on fluxes and how they are channelled, on symbiotic relations between humans and non-humans, and on the circulation of relevant signs.

A fifth suggestion is to multiply the number of other humans included in the extended body. This is a means of tackling issues of ownership and sharing, and of creating new meanings through interesting interactions. For instance, in a family, there are multiple coexisting perceptions: all human bodies in the household have different perceptions that are more or less coordinated. In this context, if the injunction to 'save energy' is to have any meaning, the extended body (the family) has to organise its parts and movements with reference to this goal. This is likely to require recurrent discussion. The recommendation to increase the number of human bodies in an extended body has other advantages. One is that it draws attention to inequalities between people. Inequalities of income, knowledge and skill are partly due to an unequal distribution of relations between humans. If humans (rather than commodities) could freely travel, meet and exchange, inequalities between extended bodies might be reduced. These encounters would create new practices and meanings. That said, infrastructures, walls and borders obviously play an important role in separating humans.

To conclude, I would like to come back to the question of whether individuals always seek to extend their bodies. The human body is at the centre of the extended body: the term 'extended' is applied to the body. Agency is distributed, but differences remain between humans and non-humans. In a sense, I have argued that an extended body expresses a way of living. We extend our bodies to find new configurations, new ways of living, that are meaningful. Living is consequently defined by the potential to experiment with new assemblages. The way we want to live is directly linked to the way we want to develop our extended bodies. Possible actions are not predetermined, but are potential combinations of new extended bodies. Bodies can be indefinitely extended, but the question – for sustainability – is about how we can prevent ourselves from becoming slaves to our own extended limbs. The ethical principle of the adequate body (in Spinoza's sense) relates not to the maximum possible extension of one's body but to the possibility of experimenting with new and meaningful assemblages. Rather than multiplying extended bodies, it is possible to intensify a selected set of geometries. Moving towards the goal of sustainability consequently depends on realising that the power of human bodies is increased not by the extension of their parts, but by the intensification of their internal relations.

Notes

1 Gould (1989) sees the pattern of evolution more in the form of a bush than a tree: the history of life is not determined. There is no royal road towards even more complexity. Gould and Basalla share the view that other histories could have been possible.
2 Following the symmetry principle, the reverse is also true: while performing the horse is equipped with a human.
3 The propagation of practices through hybridisation, for example, is closer to plant than to animal forms of reproduction.

References

Basalla, G. (1988) *The Evolution of Technology*, Cambridge: Cambridge University Press.

Bennett, J. (2010) *Vibrant Matter: A Political Ecology of Things*, Durham NC: Duke University Press.

Bennett, J. (2004) 'The force of things: steps toward an ecology of matter', *Political Theory*, 32: 347–372.

Butler, S. (1912) Lucubratio Ebria [From the *Press*, 29 July, 1865], The Note-Books of Samuel Butler, available online at: www.gutenberg.org/dirs/etext04/nbsb10h.htm#startoftext (Accessed 6.7.2012).

Collins, H. (1974) 'The TEA set: tacit knowledge and scientific networks', *Science studies*, 4: 165–186.

Deleuze, G. and Guattari, F. (1983) *Anti-Oedipus: Capitalism and Schizophrenia*, trans. R. Hurley, M. Seem and H. R. Lane, Minneapolis: University of Minnesota Press.

Deleuze, G. and Guattari, F. (2009) 'Balance-Sheet for "Desiring-Machines"' in F. Guattari (ed.) *Chaosophy: Texts and Interviews 1972–1977*, Los Angeles: Semiotext(e).

Gould, S. J. (1989) *Wonderful Life: The Burgess Shale and the Nature of History*, New York: W. W. Norton.

Gras, A. (2007) *Le choix du feu: Aux origines de la crise climatique*, Paris: Fayard.

Hallé, F. (2002) *In Praise of Plants*, trans. D. Lee, Portland, Oregon: Timber Press.

Haraway, D. (1991) 'A cyborg manifesto: science, technology, and socialist-feminism in the late twentieth century' in *Simians, Cyborgs and Women: The Reinvention of Nature*, New York; Routledge.

James, W. (1890) *The Principles of Psychology*, available online at: http://psychclassics.yorku.ca/James/Principles/ (Accessed 6.7.2012). The quotation comes from Chapter X, 'The consciousness of self'.

Latour, B. (1993) *La clef de Berlin et autres leçons d'un amateur des sciences*, Paris: La Découverte.

Latour, B. (1999) *Pandora's Hope: Essays on the Reality of Science Studies*, Cambridge, MA: Harvard University Press.

Pickering, A. (1984) *The Mangle of Practice*, Chicago: University of Chicago Press.

Raunig, G. (2010) *A Thousand Machines: A Concise Philosophy of the Machine as Social Movement*, Los Angeles: Semiotext(e).

Reckwitz, A. (2002) 'Towards a Theory of Social Practices: A Development in Culturalist Theorizing', *European Journal of Social Theory*, 5: 243–263.

Schatzki, T. (1996) *Social Practices: A Wittgensteinian Approach to Human Activity and the Social*, Cambridge: Cambridge University Press.

Shove, E. (2009) 'Everyday practice and the production and consumption of time' in E. Shove, F. Trentmann and R. Wilk (eds) *Time, Consumption and Everyday Life: Practice, Materiality and Culture*, Oxford: Berg.

Shove, E. and Pantzar, M. (2005) 'Consumers, producers and practices: understanding the invention and reinvention of Nordic Walking', *Journal of Consumer Culture*, 5 (1): 43–64.

Simondon, G. (1980) *On the Mode of Existence of Technical Objects*, trans. N. Mellamphy, University of Western Ontario, Ontario. [Originally published in 1958 as *Du mode d'existence des objets techniques*].

Spinoza, B., (1677) *Ethics*, available online at: http://frank.mtsu.edu/~rbombard/RB/Spinoza/ethica3.html#Definitions (Accessed 6.7.2012).

Warnier, J.-P. (1999) *Construire la Culture Matérielle: L'homme qui pensait avec ses doigts*, Paris: Presses Universitaires de France.

Winance, M. (2006) 'Trying out the wheelchair: the mutual shaping of people and devices through adjustment', *Science, Technology and Human Values*, 31(1): 52–72.

Part V
Sustainability, inequality and power

11 Power, sustainability and well being
An outsider's view

Andrew Sayer

Introduction

As an interested outside observer of practice approaches and issues of sustainability, I will present some critical, but hopefully constructive, observations on the research paradigm presented in this volume.

The enormity of the task of achieving sustainable ways of life has been widely noted. People in developed countries have become thoroughly accustomed and adjusted to high consumption, and high carbon, ways of life. Capitalism itself is addicted to growth. Achieving more sustainable ways of life is likely to require major changes to everyday practices (and considerable political will) as major interests are threatened. It would seem that fundamental changes in values and ways of thinking and acting are required. Although there is little evidence of such political will, what efforts there have been in the realm of government policies have tended to rely mostly on exhortation.

Yet practice theory tells us that behaviour rarely changes simply as a result of challenges to ideas, values and attitudes. So what is the place of reasons and values in practices? To ignore them would be questionable both descriptively and normatively: it would be strange to deny that they have any influence on what we do, and unethical to advocate policies that bear upon practices without considering what people involved in them think and value about them. In particular, the place of lay values is crucial in relation to any agenda of change. Seeking more sustainable ways of life implies a motive or reason for such change. But how is the view of that political goal related to the view of practices, and the values and valuations of those involved in them?

As there have been disagreements in practice theory regarding the appropriate conceptualization of people's understandings of what they are doing within practices, the first part of this contribution will discuss the place of individuals' reasons and values in what they do. The second part addresses a very different set of questions, involving a different kind of value – economic value. There is an extraordinary irony in government efforts to promote more rational and sustainable forms of behaviour in individuals in economies which are dominated by the imperative to make profit through capital accumulation, and which are inevitably addicted to unsustainable growth. Too much of the practice literature ignores political economic matters mediating practices. As Tim Jackson and others have shown, the idea of a green capitalism, in which growth of output is 'de-coupled' from greenhouse gas emissions, is an impossible dream. I shall therefore argue that practice approaches need to consider political economic matters.

Practice theory, agency and normativity

Practice theory is attractive both from an academic standpoint as an explanation of what people do and how they do it, and from a pragmatic point of view, as a source of ideas on how to effect change. On both criteria it is superior to theories of behaviour change which (a) abstract particular behaviours from the practices in which they are normally embedded, and (b) attribute their adoption to simple processes of learning and responding to information, arguments and exhortation, or to external constraints, so that motivations (and hence actions) are changed.

Most commonly, these take the form of the 'ABC model', in which Attitudes drive Behaviour, which produces Change, or in some variants, Attitudes drive Behaviour within the constraints of Context, where the latter is seen as having only external, negative effects (Shove 2010). Variants of the ABC model have been used in a wide range of cases, from health (for example, campaigns to improve diet) and road safety, to energy use and recycling. In relation to public policy, it fits with neoliberal attempts at 'responsibilizing' individuals to change behaviours that are always also the product of social relations and practices; for example, potentially 'blaming' individuals for their unemployment, while diverting attention from the role of capitalist social relations and pressures in causing it (Lemke 2001). Yet policies based on ABC have had limited success. Improved information, corrections of myths and promises of improved quality of life rarely have the impact on behaviour that might 'rationally' be expected, even where individuals accept the rightness of the messages that are being purveyed. But, then, if we consider many of the major changes in ways of doing things that we ourselves have experienced, be it from sending paper memos to sending emails or from bathing to showering, it is clear that we did not change simply through this process of learning and revising our motives and actions.

Behaviour is better understood in the context of practices which are socially and materially embedded, so that individuals are constrained, enabled, motivated and pressured into acting in ways that are consistent with them. Indeed, as Bourdieu has emphasized, people themselves are already shaped by practices, for they acquire dispositions, habits of thought and action, and templates of how to do things, that fit with the situations they have experienced (Bourdieu 1984). Unlike ABC and rational choice approaches, which treat actors' understandings out of context and as merely propositional, practice theory emphasizes practical, embodied understanding or know-how.

However, habituation to situations does not necessarily mean acceptance. People may resist or at best comply reluctantly, for example, doing an unpleasant and stressful job because they need the money; there exists a micro-politics of everyday life. Nor should we merely invert the ABC model and deny any space for more conscious deliberation. Everyday action exists along a continuum, from behaviours that arise from habituation to situations and the acquisition of appropriate practical dispositions, through to actions which depend on these dispositions and skills but also involve more or less conscious monitoring and steering and to actions that rely on more focused evaluations in which we try to resolve what to do. The more conscious deliberations are relatively rare and already influenced by the more or less continuous, and semi-conscious, 'work of attention' in the flow of everyday practices, as Iris Murdoch puts it:

> ...if we consider what the work of attention is like, how continuously it goes on, and how imperceptibly it builds up structures of value round about us, we shall not be surprised that at crucial moments of choice most of the business of choosing is already over. This does not imply that we are not free, certainly not. But it implies that the exercise of our freedom is a small piecemeal business which goes on all the time and not a grandiose leaping about unimpeded at important moments. The moral life, on this view, is something that goes on continually, not something that is switched off in between the occurrence of explicit moral choices. What happens in between such choices is indeed what is crucial.
>
> (Murdoch 1970: 36)

The relatively infrequent periods of deliberation may involve 'stepping back' from practices, away from the usual pressures; but then that very abstraction or removal can mean that people fail to grasp what they have to deal with because, like the ABC model of behaviour, they are likely to overlook their embeddedness in practices. Thus, anyone who does decide to reduce their carbon footprint significantly faces considerable problems that make their accustomed practices difficult. They may underestimate just how difficult it is not only through ignorance but because, quite simply, it is one thing to decide what to do, and another to actually do it in the midst of the onward rush of everyday practices and their pressures. Nevertheless, that they do regard their habitual behaviour as problematic is significant, even if it is not very effectual. And it is significant precisely because the world is open and things can take a turn for better or worse, so that we have to evaluate what is happening continually.

There is some variation within practice theory in how the understandings and values of actors are conceptualized. According to Schatzki, the theory acknowledges that 'practices are centrally organized around shared practical understandings.' (Schatzki 2001: 3), but it differs from approaches that '...privilege individuals, (inter)actions, language, signifying systems, the life world...' (Schatzki 2001: 3) and it tends '...to reduce the scope and ordering power of reason'. Furthermore, '...practice thought also joins other contemporary currents in undercutting individual subjects as the source of meaning and normativity (value too)' (Schatzki 2001: 5).

I want to suggest first, that this latter tendency may be mistaken and second, that it has political implications that might be of concern. In reaction to rational choice approaches implicit in the ABC model, there may be a temptation to ignore actors' understandings altogether – which leads to behaviourism, the doctrine that action can be explained in terms that exclude reference to conscious decision or internal mental constructs. Even where discourses and understandings are acknowledged, sociologists have a longstanding tendency to treat people as dupes, whether as norm-followers, bearers of roles, internalizers of discourses, interpellated subjects or products of 'subjectivation'. For example, in the latest poststructuralist incarnation, 'subjects' – shoppers, for example – are said to be 'constituted' by the practices and settings (such as shopping malls) in which they are found. Needless to say academics who present this demeaning view of others, tend not see themselves in this way. As people are not accorded much autonomy, a manipulative politics that operates not by consulting or arguing with them but by changing the forces bearing upon them is invited. Whatever the limitations of the more cognitivist approaches to behaviour change, such as the ABC model, they at least (a) acknowledge people as capable of reasoning and judgement, and (b) imply a respect for their right to decide how to respond as they see fit.

Even if we take seriously the need to attend to peoples' shared practical understandings, there are ways of doing this that still belittle agency by representing the understandings as stable and 'internalized' by actors. For example, conceptualizing individuals as 'carriers' of practices might represent them as passive, ignoring their dynamic, normative or evaluative relation to practices. A spectators' view can produce a third person account of practices in which characteristic events recur and relations are reproduced in a predictable fashion, when to the participants, what will happen is uncertain and a matter of concern. The blandness of the representation of life as the reproduction of practices and their understandings is further accentuated where the normative force of matters for people is ignored, as when, for example, a funeral is reduced to the performance of ritual, and grief is edited out as if it were insignificant (Rosaldo 1989).

With regard to both factual beliefs and values, practice theory (like most sociology) emphasizes 'horizontal' relations among people and within discourses, at the expense of the 'vertical' relation of ideas and values to the things they are about. Thus, people's evaluative relation to the world is at risk of being reduced to their values as a stable set of beliefs about what is good or bad. It is, then, easy to overlook any influence from or dependence upon what happens. Yet our relation to the world is one of concern, in which we assess how events relate to various conceptions of the good (Archer 2000; Sayer 2011, Thévènot 2001). As evaluative beings we continually monitor – sometimes just subconsciously – how the things we care about are faring. It is a matter of concern to us, for example, that our relationships to significant others are on a good footing, and that we can pursue our commitments. Such evaluation routinely goes on within practices, some of it semi-consciously as Murdoch argued, and can also extend to assessment of the practices themselves. People have conceptions of good which are differentiated according to the practices concerned, so that the standards of evaluation for a sport are different from those of care-giving. Some of these conceptions of good, or internal standards as MacIntyre calls them, may be primarily technical (such as those of science), or aesthetic (as in art), but all practices have moral standards for the simple reason that they involve people interacting in ways which inevitably impinge on their well being. Practitioners may not only share and conform to, but re-evaluate and sometimes contest, the standards and norms of practices; indeed, various aspects of them may be politicized (MacIntyre 1985).[1]

An emphasis on norms, treated as no more than conventions, has the ironic effect of de-normativizing and de-politicizing conduct, by removing any notion of active judgement, evaluation or normative force; people just follow the norms (Thévènot 2001; Sayer 2011). An emphasis on discourses can have a similar result, unless it is noticed that discourses typically contain arguments, contradictions and disagreements (not merely consistent sets of ideas) so the relation of people to them has to a certain extent to be active and evaluative, not merely reproductive. However, it is also often possible in everyday life to get by with contradictory ideas and actions because they are related to different practices (Billig 1996).

Values are 'sedimented' valuations of things (including persons, ideas, behaviours, practices etc.) that have become attitudes or dispositions, which we come to regard as justified. They merge into emotional dispositions, and inform the evaluations we make of particular things, as part of our conceptual and affective apparatus. They are more abstract than the particular concrete evaluations from which they derive and which they in turn influence. They may be associated with a particular practice (for example, medical ethics) or be common to many (e.g. valuations of virtues like kindness).

Values are not merely backward looking but forward looking, influencing actions. For example, someone who has known both respect and disrespect, and the hurt caused by the latter, may come to believe that being respectful matters a great deal. This value is thus based on repeated particular experiences and valuations of actions, but it also tends (recursively) to shape subsequent particular valuations of other people and their actions, guiding that person's own actions and sensitising them to any sign of disrespect. The acquisition and transformation of values in everyday life lies between the two extremes of passive osmosis and extended reflection on experience. Our 'professed values' may differ from our 'values in use', not necessarily through deliberate deceit, but through lack of reflection and self-knowledge. Insofar as valuation involves empirical monitoring, it is possible for unexpected outcomes to produce dissonance, which provokes us to change our values – though this tends to be a slow process requiring considerable repetition and reinforcement. Nevertheless, it is not unusual for people to change their valuations and values, and to be able to provide reasons for such changes. Reflexivity is not the preserve of academics. To be sure, we are all influenced by social norms, but they do not necessarily merely ventriloquize people, for norms themselves are influenced by people's dynamic, open-ended and evaluative relation to the world (Sayer 2011; Thévènot 2001; Zigon 2008).

In attending to this evaluative relation it is, of course, important not to sacrifice the gains of practice theory; we need to acknowledge what Thévènot calls 'the natural and artificial equipment of the human world' (Thévènot 2001: 57) in which we are embodied and embedded. As the editors put it 'people consume objects, resources and services not for their own sake but in the course of accomplishing valued social practice'. This implies the capacity of individuals to evaluate those practices – valuation implies valuers – and at least a capacity to try to act according to their valuations. To be sure, valuations do not necessarily translate into action, and changes in everyday practices often come about without individuals consciously deciding to do things differently, but rather as part of the ongoing development of material culture and social life; yet people can still evaluate them, and sometimes this may lead them to resist changes or turn them to their own ends.

If values and valuation in practices themselves have been neglected in practice research, they can hardly be ignored in policy discourses, which are, of course, more clearly normative in character. What, then, is the relation between academic research and policy – in this case, between practice theory and political agendas, such as those promoting sustainability? Here, behaviourism runs into a contradiction: on the one hand ordinary people's active, evaluative relation to what they do is not acknowledged, while on the other, policy makers' actions with respect to them are, of course, assumed to be guided by precisely such normative reason. This is problematic not only because of its inconsistency, but because of its political implications: policy and politics then become a matter of manipulation of the public by an elite, which monopolises the capacity for practical reasons. If practice theory tends towards behaviourism, then it will reproduce these problems. It can allow social scientists to present policy makers with instrumentally-useful knowledge for manipulating practices so as to achieve external policy goals, whatever they may be (green or otherwise) over the heads of those involved in the practices.[2] Again, for all its faults, at least the ABC model treats ordinary people as capable of reason. We therefore need to take Thévènot's warnings seriously.

Any attempt to change behaviour needs to engage with people's evaluations of behaviours and of the practices in which they are embedded. This is for two reasons. The first is to do with knowledge: quite simply, because actors' understandings (including evaluations) of what they do are partly constitutive of behaviours and practices, researchers and policy makers will misunderstand them if they ignore them, and thus their policies will be either ineffectual or insensitive or both. The second is to do with justice: if change is to be democratically agreed upon, rather than imposed, there is a need to engage with practitioners' valuations. This is not to imply that we should simply accept their judgements and views, for they may well be mistaken about their situation. But then, while it is one thing to refuse to grant them *epistemic* authority, that is, as having a sound understanding of their situation, it is nevertheless important to grant them *agential* authority, that is, the right to decide if they should change their practice in response to critiques of their understanding.

It is in virtue of the fact that people are evaluative beings who can contest things that, as Shove and Walker argue, practices have a political dimension, and that there is a politics to the very definition of the practices in question and who is to manage any change in them (Shove and Walker 2007). Even the concept of 'managing' practices is politically loaded as it connotes a hierarchical model – governance by an overseer rather than the practitioners themselves. Whether by management or by other means, changing practices is a thoroughly political matter, for whatever is proposed or done affects the distribution of power and people's well being. Again, that they are steered and contested from within, as well as without, implies some degree of individual reflexivity that has to be engaged with, even if people's beliefs and actions diverge. It would be peculiar to note this political character while ignoring how actors view, contest and reproduce practices. Wherever there are calls to change behaviour, there are inescapable normative issues concerning the justification of such change: why should behaviour be changed, and if it should, in what way? It is vital that academics answer this.

As I shall now argue, part of being aware of the political implications of practice approaches involves taking capitalist economic logic more seriously.

Bringing political economy back in ... or, when the bottom line is the bottom line

In view of the dependence of so many practices on energy supplies, and the dominance of that supply by major capitalist firms, we cannot expect to understand current practices or potentially more sustainable ones without considering political economic matters. However, practice approaches tend to concentrate on consumption-related activities, though they can be used for studying businesses. This emphasis on consumption suits capitalist businesses (and governments under their influence), for it is extremely threatened by the goal of sustainability; it is therefore preferable for sustainability to be seen as the prime responsibility of ordinary people. It is important that practice theorists interested in sustainability resist this narrowing of the field.

Even where businesses are examined, it is possible to study their economic practices in a way which fails to grasp the particular capitalist economic logic (in the form of the rule of exchange-value, money, cost and profit) that drives and shapes them. One can analyse business practices in detail, looking at the everyday doings and social interaction of workers – currency dealers, bond traders, or whoever – and the technologies and procedures they use. But this can easily become a study of economic practices with the economics taken out if the logic and consequences of exchange-value are overlooked (Sayer 2001; Jessop and Oosterlynk 2008). Such studies tell us little or nothing about economic (and especially macro-economic) forces, let alone why we now have a massive economic crisis (e.g. Knorr-Cetina and Bruegger 2002).

It is not only that the capitalist economy is so much bigger than particular practices – though that is part of the problem; the practices of financial traders in London are dependent on flows of money and production and consumption practices across the world. It is also that it is based on the peculiar logics of money and capital. Practices vary greatly in their degree of dependence on these, but few are entirely insulated from it; most obviously, income enables engagement in practices while poverty, debt and unemployment restrict it.

Money depends on particular practices (particularly those of production and exchange) but is also indifferent to them, operating 'without regard for persons', or indeed, meanings. Although money can be given particular meanings (as Viviana Zelizer has demonstrated) once exchanged, the money itself bears no trace of this (Zelizer 1994); one pound is one pound, whether it is spent by an arms dealer or a charity. As writers from Aristotle to Shakespeare, Marx and Simmel have noted, with varying degrees of astonishment and disapproval, it can flow through and influence diverse practices with complete indifference to their meanings or their health (Marx 1973, 1991; Simmel 1978). Money commensurates the incommensurable: a salary, a meal, an arms deal, a hospital, a loan, an air ticket.

As a measure of value, money coordinates and regulates a host of activities according to costs. Thus, a rise in the price of petrol or electricity unilaterally affects diverse practices. Such price changes are the product of myriad events and decisions across the global economy, which are largely ignorant of, and indifferent to, their impact on people and their practices. To be sure, participants in a practice can decide what to do in response to changing prices, but the price changes themselves are the product of countless interactions and their unintended consequences. The more we depend on money, the more we are subject to 'the icy water of egotistical calculation' (Marx and Engels 2002).

Further, capitalist production of goods and services is dependent not just on money but *money-capital*, i.e. money put into producing goods or services for sale at a profit and on an ever-increasing scale. To survive in the face of competition, capitalist firms have to make a competitive rate of profit. Profit is the end or goal, and the use-values produced are just a means to this end, with environmental changes a by-product. The goals of doctors may be to heal, of teachers to teach, of golfers to play golf, but where their practices are co-opted by capital, their internal goods and standards are open to corruption by a quite different logic in which the ultimate criterion is profitability. The operation of what Marx called 'the law of value' through increasingly global processes of competition (and as technologies and ways of doing things, scarcities and the balance of class forces change) create constant, yet uneven, pressures to reduce costs and increase sales. These pressures, transmitted by the continuous flows of money in and out of practices, also regulate them.

As Schumpeter's term 'creative destruction' signals, capital is very good at *stopping* practices, through disinvestment. It is the bottom line – the rate of profit – which governs the switching of investment in or out of businesses. Major capitalist firms can often change parts of material culture where they can profit from doing so, even if the change is not actually initially 'demanded' by their customers. Such changes can ripple through to a range of practices that depend on these material cultures. Firms can stop producing old technology products or old service practices, so that users who want to renew them find they are diverted instead towards newer generations of products and practices. To be sure, as studies of innovation have shown, the success or failure and mode of adoption of innovations depends on how they fit with existing practices, so that change tends to be incremental and path dependent. But we must not leave out the economic side of all this, for it produces forces of its own that drive, direct and mediate change.

Capital can also often block changes that threaten it. It is not in the interest of energy companies for customers to reduce their purchases, or to switch to greener but less profitable energy sources, though of course they may pretend otherwise for the sake of corporate image.[3] The lead times of investments in energy extraction and production are particularly long and so companies in this sector are reluctant to write them off prematurely. As Urry notes, the energy sector has been very effective in getting media and politicians to deny anthropogenic climate change, though receptiveness to their campaigns is assisted by the very inconvenience of the implications of climate change for everyday ways of life in the developed world (Urry 2011).

Energy companies are a key player in neoliberal plutocracy and ensure that governments are compliant. Take the case of BP – which temporarily identified itself as 'Beyond Petroleum' in 2000. It is the fifth largest company in the world. It constitutes 9 per cent of the capital on the London Stock Exchange and its shares form one-sixth of the invested value of pension funds in the UK,[4] so inevitably the government is likely to assume that what is good for BP is good for British pension holders. On its website it offers visitors an energy (carbon footprint) calculator. It thereby encourages them to think that reducing carbon emissions is just an issue for individuals, rather than companies, as if the latter has just passively responded to whatever consumers demanded. James Marriot calculates that the carbon footprint of BP's production and distribution processes, and its products, is twice that of the UK (Marriot 2008). At the same time, BP is massively investing in the extraction of oil from bitumen ('tar sands') in Canada, an extraordinarily energy-intensive and polluting technology. A further irony is that global warming is opening up the possibility of new fossil fuel extraction in the Arctic. The cartoon on the next page hits the nail on the head.

Of course, ordinary people do have a responsibility to address climate change but, beyond a certain level, it is very difficult for them to reduce their carbon emissions further, given their dependence on external energy systems that rely principally on fossil fuels. Responsibility lies heavily with capital, especially in the energy sector. Even radical changes in people's everyday practices will only have limited effect on carbon emissions unless ways of producing low carbon energy can be found. BP and other energy producers can be confident that insofar as domestic consumers do reduce their direct energy consumption, the reductions will be small and probably offset by increases in indirect energy

176 A. Sayer

"And so, while the end-of-the-world scenario will be rife with unimaginable horrors, we believe that the pre-end period will be filled with unprecedented opportunities for profit."

Figure 11.1 By permission of Conde Nast Publications/The Cartoon Bank.

consumption, as savings are spent on extra goods whose production is energy intensive.

At one level, the practice approach usefully counters the tendency to attribute all responsibility for what happens to individuals and their choices. However, unless it addresses not just consumption practices but the practices of businesses (acknowledging their dependence on the logic of profit) it might be said to accommodate and merely qualify, rather than challenge, the tendency to load all the responsibility onto individuals. Governments, working in the shadow of neoliberal plutocracy, are also reluctant to challenge capital. Unless a practice approach can be combined with political economic analysis, it is likely to be seen as safely depoliticizing.

We need to ask what kinds of practices could be compatible with radically reduced CO_2 emissions and assess their implications for well being. But most fundamental of all, is the question of whether such a goal is even possible for capitalism and contemporary lifestyles in rich countries. Complementing the expansionary imperative of capital accumulation is a culture of *pleonaxia* – insatiable desire for wealth and material possessions. We should not underestimate how deeply embedded this is; the need for growth is a premise shared by most of the political spectrum, indeed progress is seen as synonymous with growth. Outside a tiny minority of green thinkers, 'de-growth' is unthinkable. Capitalism cannot survive without continually growing; zero growth sends it into crisis and recession, generating unemployment and decline. And as we have seen in recent years, recessions tend to lead to calls for economic growth and push environmental concerns into the background.

The only way capitalism could be greened would be by 'de-coupling' economic growth from growth in greenhouse gas emissions. However, as Tim Jackson has demonstrated, such an outcome is highly unlikely. Energy and carbon intensities per unit of output have declined, but CO_2 from fossil fuels increased by 80 per cent from 1970 to 2007, and 40 per cent since 1990; it grew at 3 per cent per year between 2000 and 2007 (Jackson 2009: 71). Moreover, this growth occurred during a long period of stagnation of wages and salaries, hence the relatively slow growth of aggregate demand in many industrialized countries, which limited the profits of productive capital and diverted money-capital instead into financialization (Crotty 2005; Harvey 2011; Alvaredo et al. 2011). Capitalist recovery would make the possibility of carbon reduction even more remote. Frankly, capitalism is incompatible with saving the planet.

So the idea that existing, ever-rising high consumption and high mobility lifestyles might still be possible in a low carbon economy is a delusion. Sustainability will require not only greater energy efficiency and lower carbon consumption, but lower consumption in the rich countries, period; in other words, it requires *frugality* and reduced mobility. Within countries, individual carbon emissions correlate strongly with income, so it is the rich whose consumption needs to be reduced most, though that of course would require reducing their income. Since 2007, however, what have come to be known as 'austerity politics' have impacted mainly on lower- and middle-income groups, while the rich have continued to increase their share of national income. At the same time, merely redistributing this wealth to lower- and middle-income groups would do nothing to lower overall consumption and, indeed, would raise it because of the lower propensity of low-income groups to save. While there is (of course) a strong case for allowing the lowest income groups to consume more, overall, production and sales of consumer products would need to fall in rich countries and be switched into green investment if there is to be any chance of reducing greenhouse gas emissions. (I leave aside the very different situation of developing countries, for which quite different measures would be required). The basic dilemma is that unless alternative kinds of economic organization and regulation are developed, we will be stuck with a fundamental contradiction between the interests of the planet and its population, and the need of capitalism to grow regardless of such interests. The infeasibility of a green capitalism is a truly inconvenient fact not only for the rich, but also for the comfortably off in affluent countries; it would be irresponsible of academics not to broadcast this message at every opportunity.

However, one source of hope here is that beyond a certain level of income, further increases in wealth have little impact on well being. Above this threshold, social cohesiveness, friendships, health and security make more difference. As Richard Wilkinson and Kate Pickett's influential book *The Spirit Level*, and a wealth of other research on happiness and well being, shows, *pleonaxia* is not the route to well being (Wilkinson and Pickett 2009; Lane 1991). This implies that a levelling-down of the incomes of the rich and well-off need not involve a levelling-down of well being; more frugal ways of life could be good or better, though (subjectively) our attachment to high consumption lifestyles would make this hard to appreciate, at least in the short run. As Mike Hulme puts it 'The idea of climate change should be used to rethink and renegotiate our wider social goals about how and why we live on this planet' (Hulme 2009). So we might usefully try to identify practices which are both frugal and conducive to well being. In this context, it is unfortunate that many social scientists believe that well being is merely a matter of subjective 'values', lying beyond the scope of social science, rather than a state of being which we can reason about and assess.[5]

Conclusion

In conclusion then, I am calling for a practice approach which is first, more attentive to people's understandings and valuations of what they are doing, and is prepared to evaluate this. Is the aim of practice-based approaches to sustainability to provide policy makers with better ways of manipulating behaviour? Does it matter, both from a pragmatic point of view of efficacy, and an ethical standpoint of democratic involvement, whether policy engages with the active understandings of lay people? On both counts, I would suggest it does: we need to adopt an approach to practices that includes evaluation of the substantive goals or values of both ordinary practitioners and would-be external managers.

Second, practice approaches need to link to political economic analyses of the economic forces shaping practices in a way that confronts the power and responsibilities of business. They need to underline the unsustainability of capital accumulation and *pleonaxia*. As we have seen, the implied need for more frugal ways of life for the developed world need not imply a reduction in well being – though that, of course, depends on how the political struggles go.

Notes

1. Some practices are more strongly related to people's concerns and commitments than others; for example, the qualities of cycling, for an enthusiast, are important to their identity and well being, whereas for someone who only cycles occasionally to the shops, the activity is merely a means to an end.
2. This was at the heart of Habermas' critique of the reduction of reason to instrumental reason.
3. Carbon markets might seem a good idea in that they take advantage of capital's profit motive and try to steer it in a low-carbon direction; but, given the size of the energy companies relative to governments, and the fact that the implementation of such markets depends on technical knowledge and data held by the companies, the situation has been ripe for regulatory capture. Not surprisingly, they have been ineffective.
4. Marriot (2008) and www.thisismoney.co.uk/money/pensions/article-1695211/Pension-funds-sunk-by-BP-oil-spill-chaos.html.

5 Again, the capabilities approach offers one way of thinking about such 'wider social goals' (see Walker, this volume).

References

Alvaredo, F., Atkinson, T., Piketty, T. and Saez, E. (2011) *The World Top Incomes Database*, available online at: http://g-mond.parisschoolofeconomics.eu/topincomes/ (accessed June 2012).

Archer, M. (2000) *Being Human*, Cambridge: Cambridge University Press.

Archer, M. (2003) *Structure, Agency and the Internal Conversation*, Cambridge: Cambridge University Press.

Billig, M. (1996) *Arguing and Thinking: A Rhetorical Approach to Social Psychology*, Cambridge: Cambridge University Press.

Bourdieu, P. (1984) *Distinction*, London: Routledge.

Bourdieu, P. (1999) *Pascalian Meditations*, Cambridge: Polity.

Chomsky, N. (1967) 'A review of B.F. Skinner's *Verbal Behavior*', in L. Jakobovits and M. Miron (eds) *Readings in the Psychology of Language*, New Jersey: Prentice-Hall.

Crotty, J.(2005) 'The neoliberal paradox: the impact of destructive product market competition and "modern" financial markets on nonfinancial corporation performance in the neoliberal era' in G. Epstein (ed.) *Financialization and the World Economy*, Cheltenham: Edward Elgar.

Harvey, D. (2010) *The Enigma of Capital*, London: Profile Books.

Hulme, M. (2009) *Why We Disagree About Climate Change: Understanding Controversy, Inaction and Opportunity*, Cambridge and New York: Cambridge University Press.

Jackson, T. (2009) *Prosperity Without Growth: Economics for a Finite Planet*, London: Earthscan.

Jessop, B. and Oosterlynck, S. (2008) 'Cultural political economy: on making the cultural turn without falling into soft economic sociology', *Geoforum*, 39 (3): 1155–1169.

Jones, R., Pykett, J. and Whitehead, M. (2011) 'Governing temptation: changing behaviour in an age of libertarian paternalism', *Progress in Human Geography*, 35 (4): 483–501.

Knorr Cetina, K. and Bruegger, U. (2002) Inhabiting technology: the global lifeform of financial markets, *Current Sociology*, 50: 389–405.

Lane, R. (1991) *The Market Experience*, Cambridge: Cambridge University Press.

Lemke, T. (2001) 'The birth of bio-politics: Michael Foucault's lectures at the College de France on neo-liberal governmentality', *Economy and Society*, 30 (2): 190–207.

MacIntyre, A. (1985), *After Virtue: A Study in Moral Theory*, second edition, London: Duckworth.

Marriot, J. (2008) 'BP and the fuelling of Heathrow', *Soundings*, 39: 56–66.

Marx, K. (1973) *Grundrisse*, Harmondsworth: Penguin.

Marx, K. (1991) *Capital, Vol. III*, Harmondsworth: Penguin.

Marx, K. and Engels, F. (2002) *The Communist Manifesto*, London: Penguin.

Murdoch, I. (1970) *The Sovereignty of Good*, London: Routledge.

Rosaldo, R. (1989) *Culture and Truth*, London: Routledge.

Sayer, A. (2001) 'For a critical cultural political economy', *Antipode*, 33 (4): 687–708.

Sayer, A. (2011) *Why Things Matter to People: Social Science, Values and Ethical Life*, Cambridge: Cambridge University Press.

Schatzki, T. (2001) 'Introduction' in T. Schatzki, K. Knorr Cetina and E. von Savigny (eds) *The Practice Turn in Contemporary Theory*, London: Routledge.

Shove, E. (2010) 'Beyond the ABC: climate change policy and theories of social change', *Environment and Planning A*, 42 (6): 1273–1285.

Shove, E. and Walker, G. (2007) 'Caution! transitions ahead: politics, practice and transition management', *Environment and Planning A*, 39 (4): 763–770.

Shove, E., Trentmann, F. and Wilk, R. (eds) (2009) *Time, Consumption and Everyday Life: Practice, Materiality and Culture*, Oxford: Berg.

Simmel, G. (1978) *The Philosophy of Money* trans. T. Bottomore and D. Frisby, London: Routledge.

Thévenot, L. (2001) 'Pragmatic regimes governing the engagement with the world' in T. Schatzki, K. Knorr Cetina, and E. von Savigny (eds) *The Practice Turn in Contemporary Theory*, London: Routledge.

Urry, J. (2011) *Climate Change and Society*, Cambridge: Polity.

Wilkinson, R. and Pickett, K. (2009) *The Spirit Level: Why Equal Societies Almost Always Do Better*, London: Allen Lane.

Zelizer, V. (1994) *The Social Meaning of Money*, New York; Basic Books.

Zigon, J. (2008) *Morality: An Anthropological Perspective*, Oxford: Berg.

12 Inequality, sustainability and capability

Locating justice in social practice

Gordon Walker

Introduction

The strategy of conceptualising sustainable consumption and related 'demand-side' challenges of carbon reduction in terms of social practice rather than with reference to individualistic theories of behaviour change has been shown to have considerable analytical power. A substantial critique of behaviour-based analyses and policies has emerged alongside claims that theories of social practice, which better capture the shared, cultural and material basis of past and present patterns of consumption, might be mobilised by those interested in steering ongoing social change in more sustainable directions (Shove 2003; Shove 2010; Barnett *et al.* 2011; Lane and Gorman-Marray 2011; McMeekin and Southerton 2012). One problem is that the literature on sustainability and social practice has so far failed to engage sufficiently with the necessary interaction between questions of consumption, social disparity and inequality; an interaction which is particularly important for environmental and climate justice and for the fairness (or otherwise) of low carbon transitions (Agyeman and Evans 2004; Walker 2012).

This failure reflects broader limitations in recent writing on social practice, much of which tends to focus, empirically, on the 'successful' and 'skilled' performance of practice, neglecting the consequences of the 'doings' that are described, and overlooking issues of access and inclusion/exclusion. It is, for instance, hard to find examples of research that is inspired by theories of social practice, and that explicitly addresses the reproduction of abject poverty, that analyses the *failure* to successfully perform everyday practices, or that directly engages with the reproduction of social inequality and injustice. In this chapter, I want to explore the potential for developing a method of conceptualising inequality and injustice that can work with some coherence and congruity with accounts of social practice and enable us to apply normative judgement to processes of social change, including efforts to promote more sustainable ways of life.

I do not intend to provide a thorough review or analysis of the field of relevant scholarship, rather my aim is to examine the possibilities afforded by the conjunction of a specific account of social practice and an equally specific strand of justice thinking. I draw on Schatzki (2002) to provide an account of social practice, working in particular with the distinction he makes between practice-as-entity and practice-as-performance. For an account of justice, I look to the capabilities approach as developed by Sen (1999, 2009) and applied by many others (Robeyns 2006). My reason for exploring this particular combination of theories springs from the simple observation that both are concerned, at their heart, with what people do.

Schatzki defines social practice as 'a set of doings and sayings' (2002: 73). The capabilities approach understands well-being, and the space of justice, in terms of the 'doings and beings' a person realises and is able to achieve (Sen 1999: 75). This, initially, rather superficial link provides the catalyst for a discussion that begins by considering why questions of social inequality merit examination in the context of existing discussions of sustainable consumption. I then lay out the core terms of the capability approach and relate these to Schatzki's theory of social practice. I argue that questions of justice can be located in the trajectories of practice-as-entities and in related patterns of recruitment and defection, and I go on to explore the power of the capability approach through three vignettes related to 'keeping warm', viewed as a bundle of overlapping practices that have implications for sustainability and social justice. I conclude by identifying a number of directions for further theoretical interplay and development.

Why inequality and justice?

There are a number of reasons why we should consider social or socio-environmental inequalities when thinking about the resource intensity of everyday life and about whether changes in social practice might contribute to a more sustainable society:

1 *Environmental and climate justice.* In theory, justice has always been a key dimension of sustainability (issues of international and intergenerational equity figure prominently in the Brundtland report), however this aspect is rarely emphasised in policy and practice. Academics, and environment and development activists, have recurrently critiqued this absence, underlining the importance of meeting basic needs and promoting well-being as integral to strategies for sustainable development (Rauschmayer *et al.* 2011), and bringing issues of inequality and of environmental and climate justice to the fore. As first formulated in the US, accounts of environmental justice had little to say about everyday life, consumption and demand (Walker 2012). However, the globalisation of this concept, and its reinterpretation in other settings, has led to a rather more direct engagement with patterns of comparative and unequal consumption and with questions of distribution and access to resources including energy, water and food (Agyeman *et al.* 2003; Agyeman and Evans 2003; Sze and London 2008; Walker 2009).

2 *Winners and losers in sustainability transitions.* The rapidly evolving literature on transitions towards sustainability, particularly that which is centred

on (or inspired by) the multi-level perspective (Geels 2002; Elzen and Wieczorek 2005), has had much to say about processes of innovation and socio-technical change. However, it has been much less effective in conceptualising transitions in practice or in recognising the social and political implications of deliberate efforts to steer such processes (Shove and Walker 2007; Shove 2010). Movement towards a more sustainable society will inevitably produce winners and losers. This is not just a question of which supply side technologies might win. It is also a matter of which social groups stand to gain or lose as societies and social practices develop in some, but not other, directions (Walker and Shove 2007; Voß and Bornemann 2011). For example, transitions towards a lower carbon society are likely to have significantly uneven consequences: whilst some people will be able to adopt lower carbon technologies and afford higher energy prices, others will find themselves excluded, or unable to escape the effects of infrastructural lock-in. A low carbon transition can be more or less progressive or regressive, inclusive or exclusive, depending how it is pursued and through which processes, measures and interventions. It is consequently possible that a low carbon society might be characterised by more (or less) inequality than its carbon profligate predecessor. Either way, models of transition need to take more account of present inequalities and the social impact of potential outcomes. For this, they need to develop frameworks through which normative and ethical questions can be more robustly examined.

3 *Inequality, variety and social practice.* Taking social practices as the central unit of analysis has enormous power in understanding the fundamental constitution of consumption and demand, and the implications this has for resource use and sustainability. However, much of the work to-date has been concerned with explaining broad arcs of change, and analysing historical trends in the composition, character and resource intensity of the practices of which daily life is made. These narratives are at times over-generalised, capturing processes of change but paying less attention to instances in which practices remain fixed and locked-in, or in which different paths are followed. Conceptually, there is a clear expectation of variety and differentiation within categories of social practice (Warde 2005) – this is a theme for later discussion – but this understanding is not always reflected in efforts to show how practices evolve. This is important in that inequality and injustice have to do with variety and with the differential patterns of goods and bads (in a broad sense) that are reproduced in the ways that lives are lived. There are some exceptions, such as Shove's (2002: 2) discussion of mobility and inequality, which suggests that social exclusion might be understood as an inability 'to accomplish those social practices (many of which involve mobility and co-presence) required for effective social participation'. But in general, such normative questions tend to be dealt with only in passing by those interested in sustainability and social practice.

These three reasons underline the relevance of establishing a framework within which questions of social practice, and of the justice of life conditions and outcomes, can be brought together. The next section considers whether this is something that the capabilities approach might provide.

Capability and justice

Amartya Sen's work on capability and justice has been sustained for more than 30 years (for example, Sen 1982, 1999, 2009), and applied and developed in a range of contexts. The capability approach (also referred to as a 'framework' or a 'perspective') has at its core a claim as to what the appropriate 'space' (or informational focus) for determining justice should be. Sen's argument is that it is what people achieve and are able to do and be that matters when making analyses of inequality and judgements of justice and injustice. His is an *accomplishment*-based understanding of justice that 'cannot be indifferent to the lives that people can actually live' (Sen 2009: 18). The important elements of what constitutes a good and worthwhile life, the things that people value, are referred to as 'functionings'. A person's capability to achieve these functionings is where the space for determining justice is located. Functionings can take various forms:

> The concept of functionings ... reflects the various things a person may value doing or being. The valued functionings may vary from elementary ones, such as being adequately nourished and being free from avoidable disease, to very complex activities or person states such as being able to take part in the life of the community and having self respect.
>
> (Sen 1999: 75)

There is some flexibility in the framework (and in how it has been applied (Robeyns 2006)), as to whether the measure of justice is the functionings that people do (or do not) actually achieve, or whether it is their capability to achieve them. But in conceptual terms, notions of functioning and capability are clearly distinct. Sen illustrates this distinction in the over-repeated example of the well-resourced person who chooses to fast, as opposed to the famine victim who is unable to access sufficient nutritious food, and who starves. Both fail to achieve a key functioning – nourishment – but one is exercising choice, deciding to not eat despite having the capability to do so, whilst the other does not have the capability to achieve that functioning. Equalising the capability to access food thus becomes the goal, not necessarily equalising levels of nourishment.

The focus on what people do is already apparent, but becomes clearer in the distinction that Sen recurrently makes between income, as a primary good and often used indicator of inequality, and capability. Income is important in enabling people to live a good life, but is only *instrumental*: it is something that helps achieve functionings, or that supports capability. Many other things also structure capability or shape how income might be converted or translated into

Inequality, sustainability and capability

worthwhile lives that people have reason to value, including 'inborn circumstances ... as well as disparate acquired features, or the divergent effects of varying environmental surroundings' (Sen 2009: 66). As Sen states, 'the impact of income on capabilities is contingent and conditional' (1999: 88). He uses various examples to illustrate this point, including the disabled person who needs more income in order to function in various ways (such as moving around) and achieve levels of well-being equivalent to those of an able bodied person, or the pregnant woman who needs more income for nutritional support than someone who is not bearing a child. In combination, concepts of functioning and capability provide a differentiated and multidimensional perspective on how the opportunities and outcomes of everyday life are enabled and constrained.

Sen considers justice to have a plural grounding, which necessarily involves the valuing of diverse, heterogeneous objects. He argues that different features of life need to be encompassed and that human lives need to be 'seen inclusively'. But he purposefully withholds from proposing any list of key functionings on which a common foundation of justice ethics could be based, or from putting forward 'a design for how a society should be organized' (1999: 232). However, others have filled this gap, most notably Nussbaum (2002: 36) who has proposed a specific set of 'central capabilities' that 'are the most important to protect' and that constitute 'the set of basic entitlements without which no society can lay claim to justice'. These include capabilities such as 'being able to live to the end of a human life of normal length', 'being able to have good health', 'being able to live with and toward others', 'having the social basis of self respect and non humiliation' and 'being able to participate effectively in political choices'.

Much more could be said about the foundations of the capability perspective, about what it shares with other approaches to justice, and so on. But the task here is to begin to explore points of connection and compatibility with those who take social practice as the 'smallest unit' of social analysis (Reckwitz 2002: 246): it is to this that I now turn.

Capability and social practice

Sen does not use the term social practice, but variously refers to 'actual behaviour', 'actual social realisations', 'realised actuality', 'accomplishments' and 'actual opportunities of living'. Theories of practice, as summarised by Reckwitz (2002), are nonetheless alike in putting what people *do* at the centre of enquiry, in focusing on the routine reproduction of everyday life and in dealing with accomplishments and 'actual social realisations'. If we go one step further and bring the language of social practice into the capability framework, we might conclude that practices are the doings that actualise the achievement and reproduction of functionings. For example, practices of eating can achieve a set of functionings. These functionings are not limited to those of achieving adequate nutrition and sustaining health but might also include 'living effectively with others' and 'having the basis of self-respect' – functionings that in a given social and cultural setting are valued, and seen to constitute the basis (or conventions) of a normal and good life.

In parallel we might bring concepts of functioning into discussions of social practice. Schatzki (2002) sees social practices as organised nexuses of actions (doings and sayings), that hang together. For example, eating consists of many individual tasks and actions that hold together when eating is successfully performed. He argues that the 'ends' or goals of a practice, which may exist within formal rules, or be more generally found within its 'teleo-affective structure' (2002: 80) are crucial for how actions are organised and linked. Functionings can be understood as examples of such 'ends'. They are, in Sen's terms, valued forms of being and doing, which, in Schatzki's terms, explain how certain combinations of ends, actions and tasks hold together as social practices. Multiple combinations are possible. For example, a given social practice, such as eating, might contribute to the achievement of more than one functioning. Equally, interconnected 'bundles' of social practices might share common ends (Schatzki 2002: 154), and thereby work together to enable the achievement of particular functionings. This idea of bundles and of interconnected social practices will be picked up again in the next section.

Following these lines of reasoning, the notion of 'capability,' (which is crucial for the working of Sen's justice perspective), can be related to the 'putting together', that is to the integrative 'work' of social practice. To elaborate, the organisation of the doings and sayings of a social practice involves the active integration of diverse elements which Shove *et al.* (2012) arrange into the categories of *material* (things, technologies and the stuff of objects), *competence* (skills, know-how and technique) and *meaning* (symbolic meanings, ideas and aspirations). Given that integration is a creative, generative process, we might expect that some practitioners will be in a better position to enrol and integrate the materials, competences and meanings that constitute a given practice and that some will be likely to do so with more success (in terms of valued functionings) than others; in short, they have more capability. Capability to function, therefore, can be equated in some way with capability to perform or enact a practice.

To articulate this line of reasoning more carefully, we need to return to the distinction Schatzki makes between practice-as-entity and practice-as-performance. Schatzki argues that a practice exists as an entity, that is, as a pattern that endures over time and is reproduced over space – 'a temporally and spatially dispersed nexus of doings and sayings' (Schatzki 1996: 89). It is therefore possible to talk about a practice, such as eating or cooking, as an entity, so long as there is a shared social understanding of what eating or cooking involves. On the other hand, practice-as-performance refers to the recurrent enactment and reproduction of a practice by practitioners who sustain and carry it over space and time. A practice can only continue to exist if it is reproduced by practitioners who have the skill and competence to integrate the various elements (including skill and competence) of which practices-as-entities are made. Hence practices-as-entities and practices-as-performances co-exist conceptually and are mutually and recursively interdependent.

The distinction between entity and performance is helpful in identifying different points at which normative judgements might be located. As argued above, practices-as-entities exist independent of the particular circumstances in which they are reproduced. Normative judgements are often applied to practices-as-entities. Such judgements might have to do with whether a given social practice should exist at all, or in a particular form. For example, the practice of hunting foxes with dogs is one in which both the outcome and the enjoyment (that is part of the practice) have been deemed to be so ethically repugnant as to be outlawed. Although very similar equipment and skills are involved in competitive tournament shooting, and in shooting to kill, the integration of intentionality and meaning, and of rules and norms, is not at all the same. As these examples show, practices-as-entities are regularly evaluated in terms of established moral and ethical registers (such ethical and moral judgement can be seen as a form of social practice in its own right, most evidently when rooted in religious institutions). However, if we are concerned with questions of inequality and injustice it is to practice-as-performance that we need to turn.

Different understandings of performance can be brought into view when looking for links between the capability to function and the potential to enact a social practice. In the literature on social practice, there is considerable interest in how new recruits become skilled and in how they 'successfully' integrate the elements that constitute the performance of a practice (Lave and Wenger 1991; Warde 2005). This is a point at which issues of variety, inequality and differential capability come into view. Any one practice is re-enacted through a series of variously skilled performances (noting that what counts as 'success' is in itself a matter of definition within the evolving career of any given practice). This variation is not simply a consequence of the necessary processes of learning and apprenticeship that are referred to in the literature. It also relates to the uneven distribution of capability, specifically the capability to successfully integrate the elements required for an effective performance of 'the practice'. Capability might relate to income (for example in order to access necessary material elements), but also to much else that Sen would identify as structuring capabilities to function, such as one's state of health or patterns of family structure and dependence. Paying closer attention to how 'successful' performances of practice are distributed across populations, and understanding this as a reflection of differences in the capability to perform, represents one way in which theories of practice and of justice as capability might be conjoined.

A second point of engagement relates to processes of recruitment and defection that feature, although to a rather limited degree, in the literature on social practice (Shove *et al.* 2012). Once recruited, people become the carriers of a practice, reproducing and sustaining it as an entity through repeated enactment. This process of recruitment is often presented in a rather unproblematic way, perhaps due to the use of empirical examples, including hobbies and leisure pursuits, in which participation is not especially normatively charged. However, there are many situations in which patterns of recruitment are differentiated and contentious. In some cases, access is restricted in that the 'practice as entity' has embedded norms and rules, or makes certain physical or material demands, such that opportunities to participate are exclusive and particular, rather than inclusive and open to all. Examples would include strictly gendered religious practices, or legislation which excludes women from driving or from taking up certain professions, as is currently the case in a number of Middle Eastern countries. More broadly, recruitment and capability are difficult to separate. This is so in that if a potential practitioner lacks the capabilities required to practice, they are unrecruitable and excluded (at that point in time) from reproducing the practice – however willing they might be, and however actively the practice might seek to capture them. For example, being 'recruitable' to cycling as a practice (an example used by Leßmann (2011) in terms of capability and functioning) might depend on a person's ability to afford the necessary equipment, their state of health, fitness and bodily performance, the existence of a local infrastructure that permits safe cycling, and so on. In combination, these features structure the capacity to become a practising cyclist. This is something which we might (or might not) take to be a question of justice.

Similarly, we might consider the process of defection from a practice, this being the point at which someone stops being a 'carrier' (Reckwitz 2002), to be a matter of capability. For example, a practising driver may defect from driving and be recruited to public transport because their income has collapsed due to unemployment, or because their health has deteriorated, such that they lose the bodily skills required to drive. In other words their capability to continue performing as a driver has been lost and they are no longer able to reproduce this practice. Defection may occur for other reasons (maybe in order to reduce personal carbon emissions), but where it *is* a question of capability and where defection diminishes valued functionings (as when reduced mobility limits access to essential services, or to meaningful social interaction) the resulting inequality might be one that matters in terms of justice.

However, justice and capability are not only (and not distinctly) related to the reproduction of practice-as-performance, or practice-as-entity. As argued earlier there is a necessarily recursive relation between entity and performance. As cohorts of practitioners abandon a practice, and as new cohorts take it up, so the practice-as-entity is reconfigured. For example, if car driving became the preserve of a privileged subset of the population, then the characteristics of driving-as-an-entity would be transformed through the recurrent performances of its now exclusive cohort of practitioners. Driving might then reproduce inequality, and maybe injustice (a necessary distinction), in how it is understood, as well as in how it is practised. In working through these ideas, we have been able to identify forms of capability in patterns of performance (related to uneven patterns of recruitment and defection) and in the rules and norms of practices-as-entities; both of which have consequences for the valued functionings that might or might not be achieved.

Keeping warm

This discussion brings us to a point at which we can see how concepts of capability, functioning and practice may be articulated, one in relation to the other. But how might this conjunction of ideas be applied to matters of sustainability and low carbon transition? Working through a brief and admittedly superficial example helps tie this analysis back, and allows us to re-engage with questions about the relation between inequality and sustainability, these being questions with which the chapter began.

The example I use relates to energy consumption, which is deeply problematic in terms of carbon emissions and sustainability and, at the same time, deeply valued in terms of its contribution to well-being. Energy is used in providing services that in turn contribute to pleasure, communication, health, excitement, conviviality, productivity, learning and other valued qualities of life. In Sen's terms, the capability to access energy services that support valued human functionings is at the core of why access to energy matters. It is for this reason that in the UK, and increasingly in other countries too, problems of fuel or energy poverty, have been seen as matters of justice, raising questions about the basic rights and entitlements of a sufficient and healthy life (Walker and Day 2012). As Boardman states, 'everyone needs to purchase fuel to provide essential energy services, such as warmth, hot water and lighting. These are not discretionary purchases but absolute necessities' (2010: 48). Campaigners and advocacy groups have demanded that fuel poverty be acknowledged, measured, monitored and fundamentally addressed through policy responses and in their advocacy have enrolled a language of rights, for example in the UK Rights to Warmth campaign.

How can this need for realising warmth be considered in terms of social practice? It makes little analytical sense to consider 'keeping warm' as a singular social practice. Instead, warmth is better understood as an outcome (and part of the meaning or ends) of a bundle of overlapping practices-as-entities. These practices include dressing, eating and drinking and opening and shutting doors and windows (along with various other practices involving bodily movement and stillness) together with those that are more explicitly and singularly warmth-related, such as operating a heating technology or putting another log on the fire. Across this (rather open ended) bundle of diverse practices there are shared meanings or ends associated with being or becoming warm or cold, which are integrated with other meanings and ends, and with the many other elements of each of these practices.

We would expect there to be considerable diversity in how these many warmth-related practices are performed (that is, in how their particular elements are integrated together in particular performances) and in how the bundling of practices is organised over space and time. As such, we might anticipate variation, for example, in how clothing is used and combined to contribute (or otherwise) to warmth in different settings and contexts, in how heat within buildings is accessed, achieved and controlled, in patterns of eating and drinking hot food, and in how all of these are interrelated. Across this complex bundling of practices, we would expect to identify varieties of more and less successful integrations reproduced by variously skilled practitioners who nonetheless manage to keep sufficiently warm (recognising that 'sufficiency' is historically contingent and more than a matter of physiological definition (Shove 2003)). In Sen's terms, these practitioners have the necessary (but diverse forms of) capability to keep warm and to thereby contribute to the achievement of key functionings that matter in terms of justice, including bodily health (through sustaining a healthy body temperature and avoiding cold-related illnesses) and effective education (through having a comfortable environment in which to concentrate, study and learn). However we might *also* expect to find varieties and combinations of performance in which the capability to keep sufficiently warm is less robust, and in which functionings are diminished as a consequence.

Three vignettes help bring out the value of focusing on capability, and help turn from a discussion of variety to one of injustice. These cases include a homeless older man, a frugal couple and a struggling single parent. I will first outline each situation before considering issues of capability and practice.

Case 1: Malcolm. Being homeless, living on the street and having no regular income, Malcolm finds keeping warm a constant preoccupation, particularly through the winter months. Over the years he has become adept at knowing how to keep out of rain and strong winds, where flows of warm air might be found, which public buildings and parts of transport systems he can linger in during the day, how to use cardboard and multiple layers of clothing to keep warm at night, where to get hot drinks and food and where, when it gets really cold, he might be able to get a warm bed for the night. As he has become older, he has developed multiple health problems that generally worsen in colder weather. He feels the cold much more, and far more painfully, than he used to. He wishes he had been able to sort out his life many years ago.

Case 2: Bill and Rachel. Being deeply concerned about their ecological impact, Bill and Rachel live in an off-grid rural cottage they bought 10 years ago and lead a deliberately frugal home-life. They renovated the cottage to a reasonable standard, have two wood burning stoves (which they use for heating, cooking and drying washing), a solar water heater and small PV panel on the roof. They both have part-time jobs which bring in a reasonable income, but their outgoings are low. They grow most of their own food and gather wood for the stoves from their own land when out walking (as well as buying logs when they need to). They have learned how to schedule daily routines to make the best use of the various off-grid technologies on which they rely. They only heat the two downstairs rooms in the winter; they dress to keep warm indoors, make use of hot water bottles, thick duvets and blankets at night, and go to bed early when it is really cold. They live contentedly and are in generally good health.

Case 3: Susan. A single parent since her partner left her two years ago, Susan finds it hard to look after her three young children. She lives in a poorly maintained, rented flat, which has damp and rotting windows and which has electric storage and fan heaters that she moves around to warm the property. She hasn't been able to find work that fits around her need to look after the children, but receives multiple state benefits. She really struggles with her household budget; the flat gets very cold in the winter and the pre-payment meter 'eats' her money when the heating is on. The Citizens Advice Bureau worked out that she had spent 30 per cent of her income on energy bills last year. She does her best to balance the books, buying the cheapest food, going without at the end of the week and heating only some of the rooms (for as short a time as she can get away with). During the winter, the kids complain about the cold and about being stuck together in the living room – they would rather be in their separate bedrooms. She finds the cold hard to cope with during the day when she doesn't have the heating on, so she sometimes goes to the shops or the library just to keep warm. She is always anxious about the kids' health and would love to move to another flat if she could.

Across these three vignettes we can see the skilled performance of practices that contribute towards keeping warm, and we can observe much variety in both the bundling of practices and the elements involved. In their own terms and contexts, each could be described as skilled performers, reproducing routines that are habituated and well developed. However, it is plainly insufficient to describe and account for these three bundles of practices as instances of variously skilled performance: there are normative matters here, which cry out for comment and attention and which the capabilities approach to justice can help us navigate.

Of the three cases, only Bill and Rachel have the capability to keep warm enough to enable key functionings to be achieved. They are able to live contented, fulfilled and healthy lives. They keep warm in ways that reflect a deliberate frugality (Evans 2011), which means that temperatures are often low, even though they have the capability to heat their home in ways more typically associated with norms of contemporary living (such as whole house central heating). In contrast, Malcolm's performance of living on the streets clearly fails to achieve key functionings. However skilled he might be as an experienced homeless person, he lacks the capability to perform interrelated practices in a way that reproduces forms of warmth required to sustain his health and well-being. Susan's situation is more complex in that she has dependent children to care for and about. Her high energy costs are leading her into debt (which has further knock on consequences), but she is locked into a home and a heating infrastructure that she can do little to improve. In theory, she may have sufficient income to keep her family warm, but she lacks the capability to do so because of the material elements on which she depends. As a result key functionings are under threat.

We might see Malcolm's position as one that is wrong and unjust; his lack of capability diminishes key functionings in ways that are obvious and striking. Bill and Rachel keep warm in ways that many people might consider to be archaic and insufficient, but they have the possibility of living otherwise. The capabilities framework enables us make a necessary distinction between these positions. With Susan, the justice judgement is, perhaps, more finely tuned. Her home may be just as warm as Bill and Rachel's, but the means she has to use to achieve this has detrimental consequences for other functionings (beyond health) and she is in a situation of fuel poverty that many would see as wrong. All three are combining sets of practices in varied ways, but their accomplishments are not equal in terms of the capability they have to support key functionings. Variety within this patterning of practice becomes a matter of inequality and injustice once we attend to the ways in which practices are performed (and integrated) and to the practitioner's capability to do otherwise.

Questions of resource consumption have not yet featured in the discussion, but, in moving in this direction, it is clear that we need to keep issues of sustainability *and* justice in view. If we applied the metrics of carbon accounting, Malcolm would score best: he practices the lowest carbon forms of keeping warm, and to shift him out of living in the streets would significantly increase his carbon footprint. We would therefore seek to do this on grounds of social justice, but not for reasons of sustainability. Bill and Rachel's low carbon frugality is virtuous and the manner in which they keep warm meets both current criteria of sustainability and social justice. Susan and her family have the largest carbon footprint, and addressing this by improving building and heating system efficiency would be progressive in terms of sustainability, also enhancing her capability to keep warm in ways that support key functionings. In contrast, alleviating her situation by adding to her income might improve her capability to keep warm, but would probably increase her carbon footprint.

Through these examples of warmth, we can begin to see, albeit in relatively simplistic terms, (1) that there is a need to evaluate practices (as performances) in terms that address concerns of sustainability and social justice and (2) that drawing on the capability perspective allows coherent distinctions to be made, and that such distinctions depend on an understanding of everyday life of the kind that theories of social practice can provide.

Conclusions and directions

My purpose in examining the relationship between social practice (on the grounds that practices are significant for consumption and sustainability) and the capabilities approach (as formulated to address matters of social justice), has been to explore the possibility of their coherent integration. I have identified moments of connection and argued that capability can be located at various points in the interrelation between practice-as-entity and practice-as- performance: in patterns of more and less successful performance, in patterns and processes of recruitment and defection of practitioners, and in the rules, norms and material requirements of practices-as-entities, which delimit who may participate in them.

Inequality, sustainability and capability 193

The capabilities framework represents a formulation of justice theory, which is largely consistent with Schatzki's understanding of social practice, and which can provide the basis for making normative judgements about where and how practices reproduce significant inequalities and diminish or fail to support well-being. There appears to be a degree of fit and an ability to move between the languages of practice and capability, which suggests that the latter may provide a coherent basis for making evaluative judgements about the former; that the former is relevant for understanding the reproduction of the latter, and that this is relevant for consumption (for example, of energy, transport, water and food) and hence for sustainability. In this way, the integrative multidimensionality of the capability framework complements the integrative multidimensionality of how social practices are generally understood. In understanding the varying and unequal 'body/thing/understanding complexes' (Reckwitz 2002) of varieties of social practice, we simultaneously reveal varying and unequal sets of capabilities, and thereby establish the basis for an evaluative system capable of informing ethical judgement and guiding the pursuit of greater justice in everyday life.

In this way, the capability approach would seem to resolve Reckwitz's distinction between justice and 'the good life':

> Practice theory encourages us to regard the ethical problem as the question of creating and taking care of social routines, not as a question of the just, but of the 'good' life as it is expressed in certain body/understanding/thing complexes.
>
> (Reckwitz 2002: 259)

The capability approach at its heart does not see a difference between justice and the good life. Justice is only known through the lives that people are able to lead, hence taking care of social routines which constitute the reproduction of everyday life, is exactly about doing this and also about pursuing a just society.

Whilst some progress has been made towards the objective that I set out at the start of this chapter, much remains to be worked through and there are further implications to be drawn out over a wider stage. Three potential directions can be mapped out. First, there are questions of freedom and choice, which I have downplayed up to this point. Freedom to choose is central to most interpretations of the capability approach (which does require interpretation due to its 'radical under-specification' (Robeyns 2006)), but has a decidedly ambivalent status in the literature on social practice, much of which emphasises habituation, routine and the structuring of daily life (Barnett *et al.* 2011). In the capability approach, individual choice is found both in the movement from capabilities to functionings (when people have capability, they can choose to achieve or not achieve the related functioning), *and* as a vital feature of the 'good life' in its own right. As Sen (2009) states:

> In noting the nature of human lives, we have reason to be interested not only in the various things we succeed in doing, but also in the freedoms that we

> actually have to choose between different kinds of lives. The freedom to choose our lives can make a significant contribution to our well-being, but going beyond the perspective of well-being, the freedom itself may be seen as important.
>
> (Sen 2009: 18)

This emphasis on individual choice and freedom has been the focus of a robust critique of the capability perspective; however, when charged with pursuing 'methodological individualism', or with supposing that individuals make choices detached from social processes, Sen counters that the capability approach does not assume such detachment:

> When someone thinks, chooses and does something, it is, for sure, that person – and not someone else who is doing these things. But it would be hard to understand why and how he or she undertakes these activities without some comprehension of his or her social relations.
>
> (Sen 2009: 245)

It is also clear that within the capability approach, choice is not seen as always and everywhere freely exercised. Indeed, the constraint on choice for those who are disadvantaged is at the core of the distinction between capabilities and functionings. This is a key point and one that helps distinguish between the terms in which routine practices might be evaluated. Following this line of thinking, accounts of the interplay between structure and agency, such as those developed within theories of social practice (Reckwitz 2002), may well have productive contributions to make in working with and developing the capability approach. Certainly, it would be wrong to conclude that the strategy of taking social practice as the unit of analysis somehow precludes the recognition of freedoms (broadly understood) as important shared and valued objectives.

Second, there is more to say about how questions of justice are located with respect to the interdependent relation between practice-as-performance and practice-as-entity. This is an issue that resonates with established debates about the value of rights-based approaches to justice (which can be aligned to the capability approach), and the limitations of focusing on achieving rights, rather than realising the substance of those rights in practical terms (Woods 2006). To use the language of social practice, a legal right could be an element of a practice-as-entity: for example, a rule which ensures that there is a general right to participate in a particular practice. However it is only through the integration of that rule with other elements and only through the iteration of successive performances of that practice that the substance of the right is realisable. And over time, that realisation may be eroded or strengthened through the ways in which patterns and enactments of performances change and evolve, such that the right is more or less consistently reproduced. Again, we can see that an understanding of the ongoing reproduction, flux and dynamics of social practice might bring useful insight into various strands of justice thinking.

Third, I have framed this discussion with reference to the particular policy preoccupations of sustainable consumption, carbon mitigation and low carbon transition in the context of the objectives and obligations of more 'advanced' economies. However, there is clearly scope for engaging with the much broader agenda of sustainable *development*, this being an agenda in which questions of basic needs are writ large along with issues of responsibility and justice between the Global North and South (Becker and Jahn 1999). In this context, the capability approach has been advocated as a way of conceptualising (and potentially operationalising) the very basic objectives of sustainable development as these relate to the livelihoods, health and well-being of very different populations to those that are the subjects of a sustainable consumption policy (Rauschmayer *et al.* 2011). On the face of it, there is no reason why the lines of thinking developed in this chapter should not be extended in this direction. Such a move might prove to be productive on several fronts at once: prompting those who write about social practice to engage with questions of abject poverty and vulnerability, and informing policy interventions such that they enhance capabilities to function at a most fundamental level.

References

Agyeman, J. and Evans, T. (2003) 'Toward Just Sustainability in Urban Communities: Building Equity Rights with Sustainable Solutions', *Annals of the American Academy of Political and Social Science*, 590: 35–53.

Agyeman, J. and Evans, B. (2004) '"Just sustainability": The emerging discourse of environmental justice in Britain?', *Geographical Journal*, 170 (2): 155–164.

Agyeman, J., Bullard, R. and Evans, B. (eds) (2003) *Just Sustainabilities: Development in an Unequal World*, London: Earthscan.

Barnett, C., Cloke, P. and Clark, N. (eds) (2011) *Globalizing Responsibility: The Political Rationalities of Ethical Consumption*, Chichester: Wiley-Blackwell.

Becker, E. and Jahn, T. (eds) (1999) *Sustainability and the Social Sciences*, London: Zed Books.

Boardman, B. (2010) *Fixing Fuel Poverty: Challenges and Solutions*, London: Earthscan.

Elzen, B. and Wieczorek, A. (2005) 'Transitions towards sustainability through systems innovation', *Technological Forecasting and Social Change*, 7 (2): 651–661.

Evans, D. (2011) 'Thrifty, green or frugal: reflections on sustainable consumption in a changing economic climate', *Geoforum*, 42: 550–557.

Geels, F. (2002) 'Technological transitions as evolutionary reconfiguration processes: a multi-level perspective and a case study', *Research Policy*, 31: 1257–1274.

Leßman, O. (2011) 'Sustainability as a challenge to the capability approach', in F. Rauschmayer, I. Omann and J. Frühmann (eds) *Sustainable Development: Capabilities, Needs and Well Being*, London: Routledge.

Lane, R. and Gorman-Marray, A. (eds) (2011) *Material Geographies of Household Sustainability*, Farnham: Ashgate.

Lave, J. and Wenger, E. (1991) *Situated Learning: Legitimate Peripheral Participation*, Cambridge: Cambridge University Press.

McMeekin, A. and Southerton, D. (2012) 'Sustainability transitions and final consumption: practices and socio-technical systems', *Technology Analysis and Strategic Management*, 24 (4): 345–361.

Nussbaum, M. (2002) 'Capabilities as fundamental entitlements: Sen and social justice', *Feminist Economics*, 9 (2–3): 33–59.
Rauschmayer, F., Omann, I. and Frühmann, J. (eds) (2011) *Sustainable Development: Capabilities, Needs and Well Being*, Abingdon: Routledge.
Reckwitz, A. (2002) 'Toward a theory of social practices: a development in culturalist theorizing', *European Journal of Social Theory*, 5: 243–263.
Robeyns, I. (2006) 'The capability approach in practice', *Journal of Political Philosophy*, 14 (3): 351–376.
Schatzki, T. (1996) *Social Practice: A Wittgensteinian Approach to Human Activity and the Social*, New York: Cambridge University Press.
Schatzki, T. (2002) *The Site of the Social: A Philosophical Account of the Constitution of Social Life and Change*, Pennsylvania: Pennsylvania State University Press.
Sen, A. (1982) *Choice, Welfare and Measurement*, Oxford: Clarendon Press.
Sen, A. (1999) *Development as Freedom*, New York: Anchor Books.
Sen, A. (2009) *The Idea of Justice*, London: Allen Lane.
Shove, E. (2002) 'Rushing around: coordination, mobility and inequality', available online at: www.lancs.ac.uk/staff/shove/choreography/rushingaround.pdf (accessed on 8.7.2012).
Shove, E. (2003) *Comfort, Cleanliness and Convenience: the Social Organization of Normality*, Oxford: Berg.
Shove, E. (2010) 'Beyond the ABC: climate change policy and theories of social change', *Environment and Planning A*, 42 (6): 1273–1285.
Shove, E. and Walker, G. (2007) 'Caution! transitions ahead: politics, practice and transition management', *Environment and Planning A*, 39 (4): 763–770.
Shove, E., Pantzar, M. and Watson, M. (2012) *The Dynamics of Social Practice: Everyday Life and How it Changes*, London: Sage.
Sze, J. and London, J. (2008) 'Environmental justice at the crossroads', *Sociology Compass*, 2 (4): 1331–1354.
Voß, J.P. and Bornemann, B. (2011) 'The politics of reflexive governance: challenges for designing adaptive management and transition management', *Ecology and Society*, 16 (2): 9, available online at: www.ecologyandsociety.org/vol. 16/iss2/art9/ (accessed on 8.7.2012).
Walker, G. (2009) 'Globalising environmental justice: the geography and politics of frame contextualisation and evolution', *Global Social Issues*, 9 (3): 355–382.
Walker, G. (2012) *Environmental Justice: Concepts, Evidence and Politics*, Abingdon: Routledge.
Walker, G. and Day, R. (2012) 'Fuel poverty as injustice: integrating distribution, recognition and procedure in the struggle for affordable warmth', *Energy Policy*, available online at: http://dx.doi.org/10.1016/j.enpol.2012.01.044 (accessed on 8.7.2012).
Walker, G. and Shove, E. (2007) 'Ambivalence, sustainability and the governance of sociotechnical transitions', *Journal of Environmental Policy and Planning*, 9 (3/4): 213–225.
Warde, A. (2005) 'Consumption and Theories of Practice', *Journal of Consumer Culture*, 5 (2): 131–153.
Woods, K. (2006) 'What does the language of human rights bring to campaigns for environmental justice?', *Environmental Politics*, 15 (4): 57–591.

Index

Page numbers in *italics* denote tables, those in **bold** denote figures.

Aalto University School of Economics 70, 81
access 11, 94, 96, 161, 181–2, 184, 189; broadband 58; internet 57–8, 60, 63; natural 81; to raw materials 64, 187; restricted 76, 110, 188; unequal 55
accumulation 11, 38; capital 167, 176, 178
activity 24, 28, 32–6, 40, 42, 44, 51, 54–6, 174, 184, 194; bodily 79, 91; component 25; emotional 149, 152; everyday 57, 69; global 95; human 31, 39, 41, 90–2; ICT-supported 58; instrumental 22; material 78; mental 19, 34, 91; non-instrumental 28; organized 37; practical 18–19, 105; reading 69
administration 56; administrators 69, 71, 74, 77, 7981
administrative 76, 81; offices 71, 77; perspective 75, 80; practices 70; tasks 56
agency 17, 157–8, 162, 168, 170, 194; multiple-human-body 155; personal 104
Agyeman, J. 181–2
Akrich, M. 143n6
animals 147, 149, 152, 154–5
anthropology 17, 28n6; anthropological interest 103
Ashtanga yoga 6, 89–91, 94–9; workshops 93
assemblages 9, 18, 138, 148, 150, 152, 154, 156, 162; stabilised 158; transient 157
associations 23, 25, 28, 127–8
automobility 6–8, 118–24, 126, 128
autonomy 17, 105, 170; autonomous 25, 160

Barnett, C. 181, 193

Basalla, G. 147, 163n1
Bennett, J. 152, 159
bifurcation 4–5, 38, 43, 45n5
bodily repertoire 34, 3940
boundary 18, 20, 151; blurred 60, 134, 137; community 105, 108; filtering 154; objects 71, *72*, 82n1
Bourdieu, P. 17, 21, 26, 28n1 35, 41, 91, 168
Britain 6, 18, 22, 25–7, 89, 98, 100, 125–7, 175; DECC 119; DfT 121–2; financial traders 174; London 106, 125–7; London Stock Exchange 175; Rights to Warmth campaign 189; *see also* London
Brundell, B. 95, 97
bundles 4, 9, 31, 36–8; complex 42; dissolution 42–3; multiple 40; new 37–9, 41, 43; stable 41–2; sustainable 44
Butler, S. 146–8, 151

Callon, M. 138, 141
Canada 89, 175
capabilities 9, 185–91; approach 182, 184, 193–5; capabilities framework 11, 192–3
capitalist 9, 168; firms 173–4; production 134; recovery 177; societies 10
carbon 177; dioxide 90; emissions 2, 5, 90, 177, 188–9; footprint 169, 192; footprint calculator 1, 175; intensive 117; miles 99; reducing emissions 103, 175; *see also* lower carbon
carriers of practice 6, 8, 10, 19, 118, 170, 187–8
cars 35, 56, 117–20, 122, 125, 127, 129n2, 148, 160; electric 133; hybrid 43; ownership 1, 127

Index

China 121, 143n3, 143n4
Christensen, T.H. 5, 49, 51, 57, 60, 65
circulation 9, 91, 95, 97–8, 137–8, 161–2; global 6, 99; increasing 90, 92–4; public 23
Citizens Advice Bureau 191
climate change 1–2, 44, 64, 117, 119–20, 178; anthropogenic 175; seasonal 104
coalescence 4, 37–8
codification 23–5, 27
coexistence 20, 28n3, 80, 82n6 159
collective 25, 92, 103–4, 152; arrangements 105; footprints 2; frameworks 52–3; game 155; learning 7; practice 20; time structures 54–5, 57, 62; travel patterns 90; understanding 109
Collins, H. 24, 159
commitments 8, 10, 23, 26, 170, 178n1; normative 117; political 126
community 7, 103, 184; Ashtanga yoga 90; boundaries 105, 108; dynamics 111–12; living planet 89; of practice 104–11; Shaker 22; threat to 96
competence 2, 5, 11, 18, 20, 22, 24–5, **52**, 69, 79, 137, 186
competition 26–7, 118, 123–5, 174, 187; competitive relations 7; international 64
conflict 10–11, 137
consumption 3–4, 11, 18, 28, 28n4, 173, 183, 193; carbon 177; Chinese 143n3; electricity 49, 63; energy 49–51, 56–9, 62–4, 175, 189; European 1; high 167, 177–8; increasing 64; media 58; patterns 2, 148, 181; practices 174, 176; resource 5–6, 89, 192; sustainable 181–2, 195
contradiction 70, 143, 171–2, 177; contradictory 27, 142
coordination 4, 20–1, 38, 51, 54–6, 59, 63, 148, 158, 161; coordinated entity 24, 51; micro 57, 62; social 23, 25–6, 71, 72, 78; strong 28; weak 25, 28
costs 2, 80, 119, 127, 134, 173–4, 191
customers 26, 34–5, 141, 175
cyborg 150–1, 156
cycling 3, 6–8, 89, 117–19, 123–4, 127–9, 144n12, 178n1, 188; decline of 122; electric 8–9, 132, 134, 136–8, 140–2, 143n3; infrastructure 126; International Bicycle Fund 121; normal 135–8, 140; risks of 124–5

defection 8, 12, 105, 118, 126, 187; from cycling 7, 122–3; from driving 124,
127–8; patterns 129, 182; process 188, 192
Deleuze, G. 32, 38, 151–2
dematerialisation 4–5, 50, 63, 65, 69–70, 77–9, 161
Denmark 126; Statistics Denmark 60
desiring-machine 150–2
diffusion 6, 58, 60, 121
digital 78; archives 73, 80; culture 79; documentation 71; information 62; technologies 5, 57; traces 135
digitalization 70, 79
dissolution 31–2, 37–8, 40, 42–4, 45n6 105
distribution 11, 22, 33, 75, 120–1, 150, 158, 160–1, 175, 182; of competencies 140; of information 56; of performances 92; of power 173; social 97; unequal 162, 187; of vehicles 134
documents 4, 69, 71, 73, 74–5, 78, 82n3; original 73, 77; policy 133
driving 6–9, 25, 43, 58–9, 74, 119, 121–5, 127–8, 188; conduct 159; displacing 117–18
Duguid, P. 69 70, 76, 79
dynamic 26, 56, 60, 63, 99, 104, 107, 109, 140, 149, 159; accomplishment 69; amalgamation 154; character 55; community 111–12; cycling practice 119, 136–7; family 55; group 39; of practices 27, 64, 118, 124; processes 8, 128, 135, 147; relation 117, 153, 170, 172; of social practice 8, 194; of socio-technical systems 122

eating 3–4, 18, 21–2, 24, 26, 28n8, 38, 53, 59, 185–6, 189; performances 25, 27; space 107
ecological 44, 152, 190; footprint 1–2, 5; innovation 7; social ecological system 3
economics 3, 18, 173
economy 45n4, 128, 167; advanced 195; capitalist 174; formal 52; new 50; political 10 11, 17, 173
education 52, 56, 129, 190; educational institutions 23, 26
electricity 147, 152, 154; consumption 49, 63; plant 34; price rise 174
Elias, N. 18, 21
Ellegård, K. 54, 57
Elzen, B. 117, 183
emergence 17, 31, 37–9, 42–4, 56, 79, 105–6, 111, 120–1, 127, 129, 149
enactment 19, 123, 138–9, 142–3, 194; of

Index 199

practice 35, 51, 148–9; recurrent 186–7; of travel patterns 90
energy 57; consumption 2, 49–50, 189; changes 57; ICT 50–1, 55–7, 62; increase 49, 63–4; reducing 49, 175
environmental 69, 132, 177, 185; changes 174; ends 105, 112; gains 50; impact 1, 5, 49–50, 52; implications 3, 49–50; improvements 49–50; justice 181–2; positive ends 105, 112; problems 63; sustainability 89, 144n12
Erdmann, L. 50–1
Europe 24, 126, 143n3; northern 121, 127
European 1, 8; market 133; northern 122; Sociological Association Conference 28n4; sociology 17; Western 1
Evans, D. 105, 191
event 32, 107, 154; temporalspatial 33–4, 36
evolution 4, 24, 37–44, 45n6, 119, 146, 148–5, 157; co-evolution 5, 117, 146; creative 152; human 147, 158; pattern 163n1; technological 147
extended body 9, 147, 149–51, 153, 155–6, 159–62; assemblages 18, 148, 152, 154, 157–8, 162

feedback 8, 126; mechanisms 120; positive 124, 127, 129; prompt 134
Ferguson, P. 25–6
flux 149–50, 154–8, 160–2, 194
formalization 4, 23–4, 26–7, 105
fossil fuels 2, 175, 177; oil 120, 123, 175
Foucault, M. 17–18, 38, 45n7, 144n9
France 26; French cuisine 25
frugal 11, 178, 190; frugality 177, 191–2
functioning 10–11, 161, 184–5, 188–9, 193–4; desiring-machine 151; extended body 150, 152, 157, 160; key 184–5, 190–2; social practice 186; valued 184, 186, 188

gastronomy 24–7
Geels, F.W. 7, 117, 120–2, 183
Germany 126, 137, 143n1, 143n4
Giddens, A. 3, 17, 19, 28n1, 32, 35, 41, 44n1 91
global north 143n3, 195
global south 195
Gould, S.J. 163n1

Hägerstrand, T. 51, 53
Hamilton, R. 125
Haraway, D. 143, 151

health 56, 123, 144n12, 168, 174, 177, 185, 187–9, 192, 195; bodily 190; child 191; individual 117, 119
healthy 189–91; eating 22, 26
heating 3, 10, 110, 157, 190–1; home 108–9, 112; system 192; technology 189
Hilty, L.M. 4950
Hitchings, R. 67, 108
hybrid 38, 156; cars 43; entity 120–1; human-non-human 9, 136; machine 151; socio-technical 150
hybridisation 4–5, 38, 43, -4, 45n5, 134–5, 142, 147, 149, 151, 155–6, 160, 163n3

individual choice 1–3, 89, 98, 104, 117, 193–4
individualist 3, 17, 36, 104, 181
inequality 183–4, 187–9, 192; global 65; patterns of 10–11; social 181–2
information and communication technologies 5, 4951, 55–65
infrastructures 2, 8, 11, 49, 52, 56, 70, 127, 129, 134, 148, 160, 162, 188; changing 69, 117; cycling 124, 126; heating 191; internet 63; road transport 25, 97, 119–23; temporalspatial 41
injustice 11, 181, 183–4, 187–8, 190, 192
innovation 11, 26–7, 64, 118, 151, 159, 175, 183; material 5, 38; niches of 118, 127; policies 65; socio-technical 7, 55; technical 117, 121, 123, 127
institution 2, 55, 123; collective 25; educational 23, 26; religious 187; social 51; socio-technical 121
institutional 1, 53; patterns 63; processes 23, 25; time-structures 57, 62
institutionalised 26; institutionalization 80
intention 18, 53, 150, 153; intentional 32, 38; intentionality 187; unintentional 99
International Energy Agency 49
investigation 21; empirical 20, 31, 51, 105

Jackson, T. 168, 177
Japan 93, 143n3, 143n4
Jensen, J.O. 51, 56
Jois, S.K.P. 89–90, 93–5; Ashtanga Yoga Institute 95
judgement 8, 23, 69, 141, 170, 172, 184, 192; active 171; moral 10; normative 181, 187, 193
justice 9–11, 172, 184, 186–91, 193–5; environmental and climate 181–2; ethics 185; social 182, 192

200 *Index*

Knorr Cetina, K. 173
Komanoff, C. 119, 127

laboratory 133, 138
labour 138–40, 142; cheap 64; labour saving devices 161; voluntary 133, 143n5
Latour, B. 134–5, 144n10 150, 158
Lave, J. 6–7, 76, 82n4 103, 105, 187
Law, J. 139
Lefebvre, H. 45n7 80
leisure walking 6, 90–1, 93, 95–9
Levy, D. 70, 79
lifestyles 27, 176; alternative 44; high consumption 2, 177 178
London 106, 125–7; Cabinet Office 122, 125; financial traders 174; Office of National Statistics 119; Stock Exchange 175; Transport for London 126; Travel Watch 126
lower carbon 112; consumption 177; society 2, 6, 89, 100; technologies 183

MacIntyre, A. 10, 26, 171
macro-economic 173; transformation 10
Marriot, J. 175, 178n4
Martin, J.L. 26, 28n2
Marx, K. 174; Marxism 17
materials 5–6, 26, 51–2, 70, 75, 79, 82, 92, 123, 137, 150, 159, 162, 186; energy-saving 44; raw 49, 64–5
Mead, G.H. 32, 44n1 76
mental activities 19, 34, 91
micro-sociologies 18, 21; micro-sociological analysis 24
Mol, A. 144n10, 144n11
money 11, 35, 97, 123, 129, 147, 169, 173–4, 177, 191
morality 10, 149, 158

neo-liberal 18
Netherlands 126, 143n4
newcomers 6, 76, 82n4 103
nexus 19–20, 122–3, 151; organized 18, 25, 27; spatially dispersed 19, 186
Nicolini, D. 81, 82n4
novices 6, 24, 76
nutrition 22, 24; modern 27; nutritional support 185

objectives 23, 26, 104, 194–5
OECD 50
office workers 7, 71, 104, 106
oil *see* fossil fuels

older people 7, 104, 108–11
old hands 6, 103
ontogenesis 157–8
ontological 22, 37, 51; individualism 18, 36; status 21
organizations 35, 42, 44, 69, 81; charitable 89; complex 80; formal 23; practice 36, 39–41, 43; state 25; voluntary 100n2
Orlikowski, W. 69, 79
Oudshoorn, N. 144n12
Outline of a Theory of Practice 17
ownership 162; car 1, 127; home 108

Pantzar, M. 51–3, 57, 70, 79, 91, 123, 138
paper 5, 69–70; arrangements 74, 78, 80; copy 80; format 71, 76–7, 82n6; paperless office 5, 78, 80; practices *72*, 81, 82n3
paper-related practices 69–71, 78–9
performance 161, 187; of an action 32, 57; of bodily actions 34, 36, 188; competent 20; of an extended body 149, 151; individual 25; of a practice 9, 51–2, 99, 118, 149–50, 160, 187; proper 4, 105; simultaneous 58; skilled 181, 191
persistence 4, 6, 31, 37, 39–41, 45n6 118, 159
phylogenesis 157–8
Pinch, T. 76, 144n12
plants 149, 152, 154 155
pleonaxia 176, 178
policy 120, 182; discourse 1, 172; documents 133; focus 104, 195; interventions 8, 124, 195; makers 128–9, 172, 178; public 168; responses 189; sustainable 161; tools 125; travel 98
political 128, 144n10; choices 185; commitment 126; conditions 49, 64; events 42; implications 170, 172–3, 183; interest 64; interventions 125; priorities 124; will 167
political economy 10–11, 17, 168, 173; economic analysis 176, 178
politics 10, 143, 172–3; austerity 177; manipulative 170; micro-politics 169; of technology 8
power 3, 8, 10–11, 25, 34, 50, 104; battery powered 132; law 127; power and dominance 55
practice 3, 18; approach 104, 112, 159, 176, 178; communities of 6–7, 103–8, 110–12; elements of 6, 10, 92, 149–50; geometry 9, 149, 151, 153–4, 156–7, 161; sustainable 23, 91 2, 104; theory 11,

18, 21, 49, 51, 104, 149, 153, 167–70, 172, 193; turn 10, 51
practice-arrangement bundles 31, 36–40, 44
practice-as-entity 11, 18, 20, 23, 63, 91, 144n8, 182, 186–9, 192, 194
practice-as-performance 11, 18–19, 182, 186–8, 192, 194
practices 3–4, 9, 35, 38, 40, 174; business 20, 22, 173; component 25, 28; compound 18, 24–7; cooking 19–20, 22, 38; dispersed 18–22, 28n8; everyday 49, 56–7, 64, 167, 169, 172, 175, 181; integrative 19–27; multiple 12, 63, 128; relations between 7–9, 40–1, 58, 119, 123–4, 127, 129, 138, 156, 159; resource intensive 12, 67, 104, 110, 112, 148; social 1–2, 5, 11, 31, 34, 36, 51–2, 57, 90, 98, 105, 152–3, 183, 186, 193; transform 5, 118; travel-intensive 99
practitioners 5–6, 43, 51–2, 55, 59, 93, 95, 99, 100n2, 160, 171–3, 186, 188, 192; competent 24; engaged 63; groups 25–6, 28, 95, 109, 111; individual 98; normal 136, 178; recruit 118, 123–4, 117, 128–9; skilled 190
pragmatism 28n2 168, 178
praktiken 18, 21; praktik 24
praxis 17
Pred, A. 51, 53, 55
professional 76, 107; associations 28; office workers 104, 106; organizations 100n2; test 133
professionals 81, 82n4
professions 23, 188
profit making 910, 129
Pucher, J. 126

Rauschmayer, F. 182, 195
Reckwitz, A. 5–6, 18–20, 51, 91, 105, 149, 185, 188, 193–4
recruitment and defection 7–8, 105, 118, 126, 129, 182, 187–8, 192
recycling 38, 50, 168; bins 73, 75; facilities 41
regulation 23, 25, 117, 120, 177; public 64; social 26
reliance 8, 71, 106, 108
reproduction 3–4, 19, 21, 24, 107, 147, 163n3, 170, 181, 185–6, 193–4
resources 1–3, 5, 28, 112, 123, 156, 161–2, 172; access to 182; consumption 56, 89, 192; dominant 98; human body 146; intensity 3–4, 9, 90, 182–3; movement of 94; natural 3, 80, 148; philosophical 152; time 53–4; use 2, 52, 54–5, 94, 99, 148, 156, 183
responsibility 123, 173, 175–6; causal 41; issues 195; transfer 103
Rheinberger, H. 45n5 138
Robeyns, I. 182, 184, 193
Røpke, I. 5, 49–51, 56, 60, 64–6
routine 6, 22, 24, 38, 55, 69, 96, 109–10, 118, 123, 127, 138, 170, 185, 191, 194; daily 58–9, 125, 190; social 193; weekday 106
rules 4, 20–1, 23–4, 26, 34–5, 92, 155; common 37, 39; explicit 19; formal 27, 128, 186; and norms 187–8, 192; tacit 127

Sayer, A. 10–11, 100, 170–3
Schatzki, T. 3–5, 17–23, 26–7, 28n1, 28n3, 28n5, 32, 36, 51, 91–2, 122, 144n8, 150, 153, 169, 181–2, 186, 193
Sellen, A. 6970, 79
Sen, A. 9, 11, 182, 184–7, 189–90, 193–4
sharing 6–7, 20, 162; information 82n3; knowledge 110
Shove, E. 3, 5–6, 10–11, 51, 53, 57, 65, 70, 79, 91, 104, 117–18, 120, 122–3, 125, 137–8, 143, 144n8, 168, 173, 181, 183, 186–7, 190
social life 2, 19–20, 31, 33, 36–7, 39–40, 43, 100, 105, 172
sociological analysis 18, 20, 22; micro 24
socio-technical 7; hybrids 150
Solnit, R. 95–6
Southerton, D. 3, 54, 99, 181
Spinoza, B. 151, 159
stability 4, 39–44, 140–1, 144n10
standards 23–7, 80, 171, 174, 190; shared 22, 134; standardisation 9, 133
steering 8, 10, 23, 117, 128, 159, 169, 173, 178n3 181, 183
Strauss, S. 92, 94, 100
structuralism 3, 17; poststructuralist 170
supermarkets 25–6
sustainability 1–3, 8, 10–11, 69, 90, 104–5, 129, 146–9, 160, 162, 167, 173, 178, 181–3, 189, 192–3; brokering 108, 110–11; environmental 89, 114n12; organizational 82n5; promoting 106, 172

Teach Yourself books 23, 28n6
technological 5, 134; advances 39, 42, 64; artefact 121; bundles 44; change 65; environment 58; evolution 147; perspective 69; substitution 122

Index

technologies 8, 38, 69, 74, 117, 120, 123, 135–6, 139, 146–7, 159, 174, 186; communication 5, 49, 56, 155; developing 2, 4; digital 5, 49, 79; financial 173; heating 112, 189; information 71, 76, 80; measurement 134; new 5, 63; off-grid 190; old 69, 175; politics of 8; renewable energy 103; studies 138; supply side 183; transportation 97–8
teleo-affective 19, 22, 158; framing 70; structure 20, 150, 186
teleological 33, 43; structures 34–5, 39, 43
testers 9, 133, 135, 137–8, 141–2
The Climate Group 49–50
The Constitution of Society 17
The Practice Turn in Contemporary Theory 3, 17
Thévenot, L. 170–2
time-space 42, 57, 63, 91–2, 159; constraints 5, 61; divide 60; interwoven 36, 39–41, 43
Toffler, A. 79–80
topographies 9, 125, 149, 153–7, 161; animal 162; different 153, 156, 160; flux 149–50, 154–8, 160–2, 194; intertwining 156; relevant 158; vegetal 155–7
traffic 25, 142; city 138–9; congestion 60; infrastructure 121; internet 63; lights 74, 129n2; systems 120
trajectories 4, 10, 99, 117, 122, 182
transformation 3, 12, 38–9, 70, 99, 118, 149, 171; macro-economic 10; process 147; systemic 7, 122, 129
transition 4, 11, 39, 103, 120; low-carbon 183, 189, 195; path 128; systemic 1, 7, 10, 117–19, 122, 129
transport 52, 57, 193; cheap 64; emissions 119; modes of 8, 89, 97, 143n4; motorised 120; personal 126; public 188; reducing 61; related 63; systems 2, 50, 190; transportation 95, 97–8, 134; urban 128, 134; utility 121–122
travel 6, 60, 96; international 99; more 61; patterns 90; personal 95; practice-mediated 98–9; rail 97–8; reduced 63, 65, 89; time 59; virtual 61, 64; walking 97

travelling 27, 61, 93, 98, 118, 121, 125
Tsing, A. 144n10

understandings 4, 11, 19–20; general 34–5, 37, 39; practical 34, 37, 39–41, 43, 169–70; shared 22
United States 38, 42, 93, 143n3, 182
unsustainability 11, 178
Urry, J. 6, 61, 96, 120–2, 175

vegetal 155–8, 160, 162
vegetarian 1; vegetarianism 162
velo-chic 127–8
velomobility 7–8, 118–26, 128–9
Voß, J. 23, 183
voluntary 90; events 32; labour 133, 143n5; organizations 100n2

wages and salaries 65, 79, 174, 177
Walker, G. 9–11, 103, 173, 179n5, 181–3, 189
Wallace, A.D. 96–7
Wallenborn, G. 9–11
Warde, A. 1, 3–5, 18, 25–6, 51, 183, 187
Warnier, J.-P. 159
wasteful 108, 161
waste management 10, 49, 82n5
water 3, 104, 146–8, 152, 154, 162, 174, 182, 189, 193; heating 154, 190
Watson, M. 78, 117
Weber, C.L. 18, 63
wellbeing 2–3, 9, 11–12, 171, 173, 176–8, 178n1, 182, 185, 189, 191, 193–5
Wenger, E. 6–7, 76, 82n4, 103, 105, 108, 187
Western world 27
Wilk, R. 53, 55
Wilkinson, R. 177–8
Williams, D. 93
winners and losers 10, 99, 182–3
Woodcock, J. 117, 119, 127
Wordsworth, W. 95–6
working life 5, 82n2 108
workplace 44, 54–5, 59–60; energy-consuming 49; tools 76
World Wildlife Fund 1, 119; Canada 89

Taylor & Francis
eBooks
FOR LIBRARIES

ORDER YOUR FREE 30 DAY INSTITUTIONAL TRIAL TODAY!

Over 23,000 eBook titles in the Humanities, Social Sciences, STM and Law from some of the world's leading imprints.

Choose from a range of subject packages or create your own!

Benefits for you
- Free MARC records
- COUNTER-compliant usage statistics
- Flexible purchase and pricing options

Benefits for your user
- Off-site, anytime access via Athens or referring URL
- Print or copy pages or chapters
- Full content search
- Bookmark, highlight and annotate text
- Access to thousands of pages of quality research at the click of a button

For more information, pricing enquiries or to order a free trial, contact your local online sales team.

UK and Rest of World: online.sales@tandf.co.uk
US, Canada and Latin America:
e-reference@taylorandfrancis.com

www.ebooksubscriptions.com

ALPSP Award for BEST eBOOK PUBLISHER 2009 Finalist

Taylor & Francis eBooks
Taylor & Francis Group

A flexible and dynamic resource for teaching, learning and research.